Frequency Synthesis by Phase Lock

Frequency Synthesis by Phase Lock

WILLIAM F. EGAN, Ph.D.

Senior Engineering Specialist
GTE Products Corporation
Lecturer in Electrical Engineering
University of Santa Clara

A WILEY-INTERSCIENCE PUBLICATION

JOHN WILEY & SONS

New York · Chichester · Brisbane · Toronto

Library of Congress Cataloging in Publication Data:

Egan, William F
    Frequency synthesis by phase lock.

    "A Wiley-Interscience publication."
    Bibliography: p.
    Includes index.
    1. Frequency synthesizers. 2. Phase-locked loops.
I. Title.

TK7872.F73E32   621.3815'36   80-16917
ISBN 0-471-08202-3

Printed in the United States of America

10 9 8 7 6 5 4 3 2 1

*To Mary and
Bill, Dan, Tom,
John, and Mike*

Preface

Applications for frequency synthesis techniques appear to be continually increasing. Synthesizers are used in CB radios, hi-fi receivers, and other communication and test equipment, where they permit with great ease and accuracy the selection of one operating frequency from many. Perhaps the primary trends accounting for this popularity are the crowding of the frequency spectrum used for communication and the increasing use of computers and microprocessors. Spectrum crowding calls both for a high degree of accuracy in transmitted frequencies, in order that channels may be closely spaced, and the easy selectability of frequencies, so that available channels can be used effectively, while computer or microprocessor control requires that frequencies be selected in response to a digital command. Synthesizers are well suited to fill these requirements. In addition, anyone who has used both modern test equipment, employing frequency synthesis techniques, and the earlier versions will appreciate another aspect of synthesis, its convenience. However, it would be difficult to meet the above requirements if it were not for the concurrent development of integrated circuits. And, along with the development of components, has come the development of many of the design and analysis techniques that are the subject of this book.

Because the technology involved in synthesizer design includes such a vast range of subject areas, the scope of this book is limited to those concepts that are peculiar to synthesizer design and to demonstrating how the various technological areas are applied in synthesizer design. Thus an entire chapter is devoted to phase detectors, which are critical components, and another is devoted to the types of frequency dividers that are almost peculiar to synthesizers. On the other hand, while the parameters of oscillators that are important in synthesizer design are given considerable attention, basic oscillator design is not. And, while the applications of control-system analysis techniques, primarily the Bode plot, are described in some detail, the fundamentals of these techniques are not taught in this book.

Within the subject area of synthesizers, the phase-locked type is strongly emphasized, but the other types, direct and digital, are not ignored. Their operating principles are described and fundamentals applicable to their design and use are discussed, but phase-lock techniques, which are probably the most commonly used, are emphasized.

Extensive references have been provided to lead the reader to a better understanding of related material that is not covered in detail, or where a more extensive, or simply alternate, treatment may be available. The footnotes usually consist of the first part of an entry that is listed alphabetically in the bibliography.

About 40 examples have been included to clarify the concepts. In addition, there are approximately 70 problems at the ends of the chapters. Many of these are similar to the examples, a factor which can be an aid to independent study. The book should be useful for independent study, as a course text, or as a reference. The minimum recommended background level is that of a graduate electrical engineer or an upper-level engineering student who has completed courses in Fourier and LaPlace transforms, digital circuits, and control systems. It should be useful not only to the designer, or potential designer, of synthesizers, but also to systems engineers who use synthesizers. Many of the concepts presented, for example, the measurement and specification of spectral purity and the effects of noise in nonlinear elements, are of value even when no synthesizer is involved in a system.

This book is intended to give a unified presentation of the most important fundamentals in frequency synthesis. It begins with a discussion of the basic manipulations of frequency in synthesis, that is, how frequencies may be added, subtracted, multiplied, and divided. Chapter 2 describes how these manipulations are employed in the three basic synthesizer types and Chapter 3 describes, in more detail, the elementary phase-locked synthesizer.

Having considered, in the first three chapters, how we might "get it to work," we then begin our study of the process of design to meet system requirements. We begin, in Chapter 4, by considering the noise sidebands that exist on all oscillators, how they and other undesired components are described, and how they are transmitted through synthesizers and through systems using synthesizers.

All of Chapter 5 is devoted to phase detectors. Six types are considered and compared. The effects of sampling on the loop dynamics and on the translation of noise frequencies are also discussed. These are rarely discussed in phase-locked-loop literature, but are very important in phase-locked synthesizers. Chapter 6 discusses frequency dividers, including prescalars, pulse swallowers, and the practical architecture of variable dividers for phase-locked synthesizers. Methods for obtaining a desired relationship between the digital control word and the synthesized frequency are given and critical timing parameters are discussed.

The more complex configurations that are often met in practice, namely, second- and third-order loops, loops with internal frequency conversion, and the Digiphase® and fractional-N techniques, are considered in Chapter 7. The second-order loop with an integrator or a low-pass or lag-lead filter is covered in a unified fashion. Properties are compared as a function of loop parameters, by means of a table of formulas, by a study of Bode plots, by equivalent circuits, and by a comparison of transient responses. The application of this theory to practical loops that do not fit exactly the second-order description is also considered.

Chapter 8 covers the nonlinear performance of the synthesizer loop during acquisition of lock, beginning with an examination of pull-in in a simple loop. The possibility of loop oscillations of limited amplitude is demonstrated here. Numerous formulas are given for second-order systems, with three different phase-detector types. One of the most important problems in phase-locked synthesizers is false lock. The most common kind is associated with the sampled nature of the loop and is peculiar to synthesizer-type phase-locked loops. This kind of false lock, as well as false lock due to excess phase shift, are discussed. Computer simulation of acquisition is illustrated and testing for acquisition is discussed. Chapter 9 describes numerous acquisition-aiding techniques to bring about phase lock when it otherwise might not occur, including one that can also improve spectral purity.

Finally, spectral purity, how to obtain it, how to specify it, how to measure it, and how various measures of spectral purity can be related to each other and to the specifications that may be used in design are discussed in Chapter 10.

WILLIAM F. EGAN

Cupertino, California
May 1980

Acknowledgment

I would like to thank the many people at GTE Sylvania who have helped to make this book possible: colleagues who have contributed to my understanding of the subject, managers who have given encouragement and help in the development and finalizing of the work, especially James Peoples and Jerry Everman, and secretaries who have typed the manuscripts, especially Dee Walraven and Connie Harper. I also appreciate the comments of my students and reviewers.

W.F.E.

Contents

Symbols

(Some symbols have more than one meaning and all meanings may not be given below, but usage should be apparent from context.)

C	Control number.
c	Subscript indicating a quantity expressed in cycle units.
D-A	Digital-to-analog converter.
dBc	Decibels referenced to carrier, that is referenced to the power in the central part of the spectrum.
E_G	Relative gain error.
E_n	Relative frequency error in the nth period.
η	Efficiency.
F	Final count; actual frequency (as opposed to f which may be a deviation from steady state).
f_I	IF, the frequency out of a mixer.
f_L	Local-oscillator frequency.
f_m	Modulation frequency.
f_{osc}	Oscillator frequency.
f_p	Filter pole frequency.
f_{REF}	Reference frequency (see Fig. 3.7a).
f_s	Sample frequency; frequency of feedback signal at phase detector (see Fig. 8.2).
f_S	Signal frequency; the frequency of the weaker input to a mixer.
f_z	Filter zero frequency.
f_0	See ω_0 below.
f_1	Open-loop frequency change at VCO output (see Fig. 3.7a).
G	Forward loop gain; gain; conductance.
G_R	Relative gain.

H	Reverse loop gain.
IF	Intermediate frequency. The output signal from a mixer or the frequency of that output.
Im	Imaginary part of.
K_F	Forward gain (in $\sec^{-1}$).
K_{LF}	Loop-filter gain constant (dimensionless).
K_φ	Phase-detector gain constant (in V/cycle or V/rad).
K_v	VCO gain constant [in Hz/V or (rad/sec)/V].
$\mathcal{L}(f_m)$	Relative SSB level at f_m from spectral center.
LO	Local oscillator: the normally stronger input to a mixer.
LSB	Least-significant bit. 3 LSB means third-least-significant bit.
m	Modulation index = peak phase deviation in radians.
MSB	Most-significant bit.
O	Offset.
ω	Radian frequency.
ω_H	Peak hold-in frequency ($\frac{1}{2}$ total hold-in range).
ω_L	Frequency where loop gain is unity. This is close to the 3-dB closed-loop bandwidth for a well-damped loop.
ω_n	Natural frequency.
ω_{osc}	Oscillator frequency.
ω_p	Filter pole frequency.
ω_{PI}	Peak pull-in frequency ($\frac{1}{2}$ total pull-in range).
ω_s	Sample frequency, frequency of feedback signal at phase detector (see Fig. 8.2).
ω_S	Peak seize frequency or lock-in frequency ($\frac{1}{2}$ total range).
ω_z	Filter zero frequency.
ω_0	Loop velocity constant. For type-1 loop, frequency where open-loop gain would be unity if loop filter maintained its low-frequency gain. For type-2 loop, ω_0 is infinite.
P	Preset number; power.
p	One-sided power spectral density.
PD	Phase detector.
PM	Phase modulation.
φ	Phase.
r	Subscript indicating a quantity expressed in radians.

Re	Real part of.
RF	Radio frequency: The normally weaker input to a mixer or the frequency of that input.
s	Laplace complex frequency.
S_φ	Power spectral density of phase.
$S_{\dot\varphi}$	Power spectral density of frequency.
sec	Second
SSB	Single sideband.
T_A	Count period (averaging time).
T_s	Sample period.
T_S	Time between starts of counts; lock or seize time.
VCO	Voltage-controlled oscillator.
VMP	Variable-modulus prescalar.
z	Z-transform variable.
ζ	Damping factor.
$\sim$	(Superscript) indicates rms.
$X\|_y$	The variable X with the condition y or referring to y.
$X\|_{y1}^{y2}$	The variable X with y between $y1$ and $y2$.
$\triangleq$	Is defined as.
$\angle x$	Angle or phase of x.
$\approx$	Low-pass filter.
$\approxeq$	Band-pass filter.

Frequency Synthesis by Phase Lock

1
Introduction

Frequency is ordinarily considered as a parameter of some variable, a variable such as voltage or current that can be used in defining the state of a system. But, in the study of frequency synthesis, a frequency can itself be considered a state variable. As with voltages and currents, frequencies can be added and subtracted in electronic circuits, and can be multiplied or divided by a constant. A frequency can also be transformed to or from other variables. Since these are the basic processes that are used in frequency synthesis, they will be discussed in this first chapter. First, however, let us briefly consider what is meant by frequency and by its synthesis.

1.1 MEANING OF FREQUENCY

If a signal is periodic over a given interval, we say that it has an instantaneous frequency defined in that interval as the reciprocal of the time between repetitions. If the period is continously changing, we can still often identify an instantaneous frequency. If the signal is considered a constant-amplitude sine wave, there is a correspondence between instantaneous magnitude and phase, and frequency is the time derivative of phase. Figure 1.1 shows a segment of a sine wave whose instantaneous frequency is f_1. A true single frequency, in the sense of a Fourier transform, refers to a sine wave with infinite duration, both before and after t_1. A sinusoidal burst of finite duration has a Fourier spectrum such as shown in Fig. 1.2, but as long as the energy is concentrated in a spectral width which is narrow compared to the resolution of interest, and as long as we are interested in times well contained within the duration of the wave, we can consider the energy to be at a single frequency.

While some processes are best understood in terms of instantaneous frequency, one must still be aware of effects due to frequency responses that are described in Fourier terms. For example, filters are generally described in Fourier terms, so their effect on a signal must be determined by considering

1

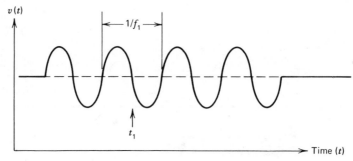

Figure 1.1   A signal which is periodic over an interval.

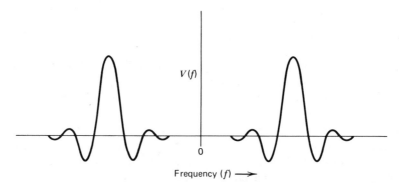

Figure 1.2   Spectrum of an RF pulse.

the signal in that sense also. We can only know whether the filter will distort a sinusoidal pulse by considering the Fourier spectrum of that pulse in relation to the filter passband.

The concept of an analytic signal[1] defines instantaneous frequency more precisely, but we will not find it necessary to go to that depth for the purposes of this book.

1.2   MEANING OF FREQUENCY SYNTHESIS

Frequency synthesis is the generation of a frequency or frequencies which are exact multiples of a reference frequency. Usually the reference is very precise and the synthesized frequencies are selectable over some range of whole-number multiples of a submultiple of the reference; that is, the output frequency will be given by

$$f_{\text{OUT}} = \frac{n f_{\text{REF}}}{M},\tag{1.1}$$

where n and M are integers, n varies from $N_{\min}$ to $N_{\max}$, and M is constant.

---

1   Bracewell, pp. 268–271.

1.3 TRANSFORMATION TO AND FROM VOLTAGE OR CURRENT

Figure 1.3 illustrates conversions from frequency to voltage, and vice versa, by means of electronic circuits, represented by transfer functions A through D. The transfer functions give the ratio of the change in output to change in input for the blocks.

The variables that we will deal with (e.g., f_1, v_1, and φ_1 in Fig. 1.3) will often represent changes, usually from an initial or steady-state value. At times, however, we will use capital F to emphasize that the actual frequency, rather than a change, is being considered.

Frequency is converted to voltage by a frequency discriminator.[2,3] A discriminator may employ tuned circuits to obtain a signal whose magnitude is frequency dependent. This is rectified to give a voltage which changes proportionally with frequency. Alternately, the discriminator may generate constant-width pulses in synchronism with the signal and take the average voltage of these pulses as a measure of frequency, as shown in Fig. 1.4. If, at some operating point, the output of a discriminator increases by 2 mV for each 1-Hz change in input frequency, then the transfer function A in Fig. 1.3 would be a gain constant equal to 2×10^{-3} V/Hz.

Voltage is converted to frequency by a voltage-controlled oscillator (VCO) wherein the tuning voltage changes the bias on a varactor diode and thereby alters its capacity and thus the oscillator frequency. Alternately, current can control frequency by modifying the magnetic field in which an yttrium iron garnet (YIG) sphere is immersed when the sphere forms part of the resonant circuit of an oscillator. If such an oscillator's frequency should decrease by 3

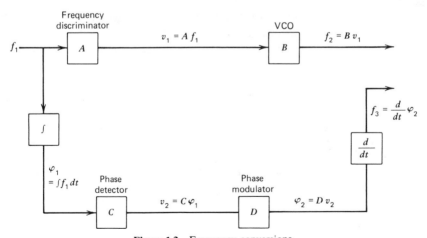

Figure 1.3   Frequency conversions.

2   Manassewitsch, pp. 407–412.
3   Klapper, pp. 37 and 186–190.

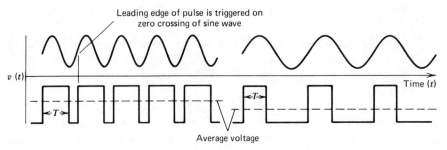

Figure 1.4   Constant-pulse-width discriminator waveforms.

MHz for each volt of increase in tuning voltage, then B in Fig. 1.3 would equal -3×10^6 Hz/V.

Phase, the integral of frequency, is converted to voltage by a phase detector. One common phase detector consists of a balanced mixer operating with its two input signals in quadrature. Both the VCO and the phase detector will be discussed further in Chapter 3, and all of Chapter 5 is devoted to phase detectors.

Voltage may be converted to phase with a phase modulator. A varactor tuned circuit can alter phase as its center frequency or its cuttoff frequency is modified as a result of a change in varactor bias. Alternately, signals which are in phase quadrature may be linearly combined to produce a desired phase. A third method is illustrated in Fig. 1.5. Here, a phase-locked loop (PLL) is employed to convert voltage to phase. Since the phase-detector (PD) output voltage is proportional to phase, and since the loop forces this voltage to cancel V_{mod}, the phase of the VCO tracks V_{mod}.

The frequency discriminator and some phase-detector circuits can be sensitive to amplitude also and may, therefore, be preceded by limiters in order to make them ideal. Similarly, some phase modulators can produce

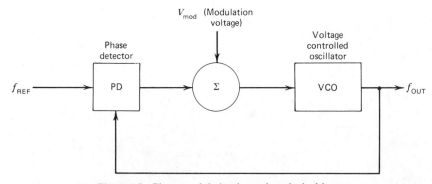

Figure 1.5   Phase modulation by a phase-locked loop.

amplitude changes. Unintentional amplitude modulation (AM) to phase modulation (PM) and PM-to-AM conversions are possible in various circuits.

EXAMPLE 1.1   The purpose of this example is to demonstrate the use of transfer functions.

Problem   For the system in Fig. 1.6, compute (a) the rms voltage at point A, (b) the average frequency at point B, and (c) the rms voltage at point C.

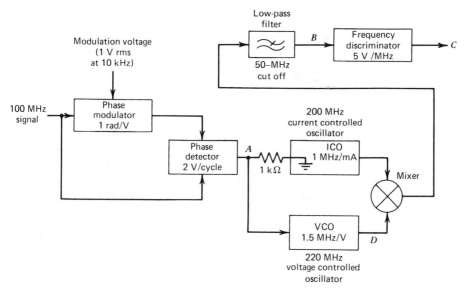

Figure 1.6   The system for Example 1.1.

Solution   (a)   The voltage at A is

$$\tilde{V}_A = (1\,\text{Vrms})\left(\frac{1\,\text{rad}}{V}\right)\left(\frac{1\,\text{c}}{2\pi\,\text{rad}}\right)\left(\frac{2\,\text{V}}{\text{c}}\right) = \frac{1}{\pi}\ \text{V rms} = 0.32\ \text{V rms.}$$

(b)   The average frequency at B is

$$F_B = 220\ \text{MHz} - 200\ \text{MHz} = 20\ \text{MHz.}$$

(c)   If the outputs from the two oscillators are $F_1 + \delta f_1$ and $F_2 + \delta f_2$, then the sum frequency is $F_1 + F_2 + \delta f_1 + \delta f_2$, which is rejected by the filter. The difference frequency passes through the filter and is $F_1 - F_2 + \delta f_1 - \delta f_2$. This produces a voltage at C of

$$\tilde{V}_C = (\delta f_1 - \delta f_2)\left(\frac{5\,\text{V}}{\text{MHz}}\right)$$

$$= \frac{1}{\pi}\ \text{V rms}\left(\frac{1.5\,\text{MHz}}{V} - 1\frac{\text{MHz}}{\text{mA}}\,\frac{1}{1\,\text{k}\Omega}\right)\left(\frac{5\text{V}}{\text{MHz}}\right)$$

$$= \frac{2.5\ \text{V rms}}{\pi} = 0.80\ \text{V rms.}$$

1.4 MATHEMATICAL OPERATIONS ON FREQUENCY

1.4.1 Addition and Subtraction—The Mixer

The mixer is used to obtain frequencies which are sums and differences of two input frequencies (see Fig. 1.7).[4,5] In the mixer, the two mixed signals* exist simultaneously in nonlinear devices (diodes). The nonlinearity produces signals with the desired sum or difference frequency, but it also produces many other signals, which can cause problems. In order to understand which frequencies are produced and how the magnitudes of the various generated signals depend on the input signals, we consider the various terms in the voltage that is produced when a current, consisting of two cosinusoids, passes through a nonlinear resistance.

The voltage may be expanded in a Taylor series as

$$v = a_0 + a_1 i + a_2 i^2 + \cdots + a_k i^K + \cdots, \tag{1.2}$$

where

$$i = A \cos(\omega_\alpha t + \theta_\alpha) - B \cos(\omega_\beta t + \theta_\beta). \tag{1.3}$$

Substituting Eq. 1.3 into the Kth-order term of Eq. 1.2, we obtain terms of the form

$$\gamma A^N B^M \cos^N \alpha \cos^M \beta, \tag{1.4}$$

where

$$\alpha = \omega_\alpha t + \theta_\alpha, \tag{1.5}$$

$$\beta = \omega_\beta t + \theta_\beta, \tag{1.6}$$

and

$$N + M = K. \tag{1.7}$$

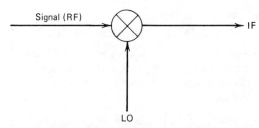

Figure 1.7 Representation of a mixer.

---

*The term "signal" refers particularly to the weaker input to a mixer but is also used in its more general sense. The weaker signal is also often called the RF input.

4   Kroupa, pp. 89–121.
5   Manassewitsch, pp. 320–337.

Using the identity

$$\cos x \cos y = \tfrac{1}{2} [\cos(x + y) + \cos(x - y)],$$

Eq. (1.4) may be expanded into terms of the form

$$A^N B^M C_{n,m} \cos(n\alpha + m\beta), \tag{1.8}$$

where

$$n \underline{\Delta} |n| = N, N - 2, \ldots, 0 \tag{1.9}$$

and

$$m \underline{\Delta} |m| = M, M - 2, \ldots, 0. \tag{1.10}$$

Usually the desired term has $n = m = 1$, and other combinations produce undesired, or spurious, signals called $n \times m$ spurs. As can be seen from Eqs. (1.8)–(1.10), signals at frequency $n\omega_\alpha + m\omega_\beta$ can be produced by terms in Eq. (1.2) of order $n + m + 2q$, where

$$q = 0, 1, 2, \cdots . \tag{1.11}$$

These combine to produce a coefficient

$$A^n \big(B^m D_{n,m} + B^{m+2} D_{n,(m+2)} + \cdots \big)$$
$$+ A^{n+2} \big(B^m D_{(n+2),m} + B^{m+2} D_{(n+2),(m+2)} + \cdots \big)$$
$$+ A^{n+4} \big(B^m D_{(n+4),m} + B^{m+2} D_{(n+4),(m+2)} + \cdots \big)$$
$$+ \cdots, \tag{1.12}$$

but if the signal at frequency ω_α is weak enough (i.e., A is small enough), the lowest-power multiplier A^n will predominate. Then the $n \times m$ spur will be of the form

$$A^n d_{n,m} \cos(n\alpha + m\beta) \tag{1.13}$$

which is proportional to the nth power of the signal strength. In dB, the spur changes n dB for each dB change in the signal and the ratio of the $n \times m$ spur to the desired 1×1 product changes by $n - 1$ dB. Thus, if the signal drops from -20 dBm to -30 dBm we expect the level of a 3×5 spur to drop 30 dB, but this corresponds to a spurious-to-signal-strength-ratio change of -20 dB. This concept is useful in extending information on spur levels from measured points.

When a mixer is used, it is generally necessary to identify the low-order spur frequencies which exist at or near desired output frequencies.[6,7] We know that a given IF frequency F_I may be produced by an LO frequency F_L and a signal frequency F_S if

$$n F_S + m F_L = F_I, \tag{1.14}$$

6 Mannassewitsch, pp. 50–54.
7 Kroupa, pp. 109–120.

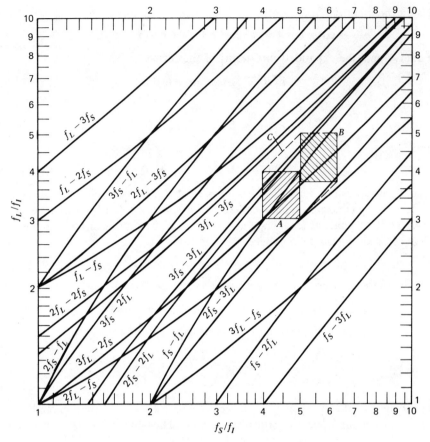

Figure 1.8 Operating regions on a spur chart.

where n and m are integers. This equation may be normalized several ways, for example,

$$\frac{F_I}{F_L} - n\frac{F_S}{F_L} = m \tag{1.15}$$

or

$$n\frac{F_S}{F_I} + m\frac{F_L}{F_I} = 1. \tag{1.16}$$

Either formula can be used to create a "spur chart" to help identify troublesome spurs. Figure 1.8 shows a spur chart, according to the last equations, with logarithmic scales. The curves show the locus of possible spurs. To identify an area of interest, the range of signal frequencies, both desired and otherwise, and of LO frequencies which can simultaneously exist, are plotted as a rectangle as shown at A. If the IF frequency changes from F_{I1} to F_{I2}, the locus moves along both axes by log F_{I2}/F_{I1}, and thus at a 45° angle. Loci for the same range of signal and LO frequencies, but a different IF, are shown at B. The entire outlined region (C) is of interest when the IF bandwidth extends from F_{I1} to F_{I2}. If any spurious frequency passes through the area of interest, then its magnitude must be determined and compared to the allowable levels.

EXAMPLE 1.2

Problem A frequency range of 100–125 MHz is converted to 25 MHz by an LO that is lower in frequency than the signal. (a) What is the range of LO frequencies? (b) Assuming a signal-filter passband that equals the signal frequency range, plot the region of interest on the spur chart. (c) Plot the region if the IF filter passes frequencies from 20 to 25 MHz. (d) What spurious responses will be produced in the IF passband by the signals in the signal range as the LO is tuned over its range? (e) If the signal strength is reduced 10 dB, how much reduction in the strength of each spurious response is expected?

Solution

(a) $F_L = F_S - 25$ MHz
$$= (100 \text{ to } 125) \text{ MHz} - 25 \text{ MHz}$$
$$= (75 \text{ to } 100) \text{ MHz}.$$

(b) At $F_I = 25$ MHz
$$\frac{F_S}{F_I} = \frac{(100 \text{ to } 125) \text{ MHz}}{25 \text{ MHz}} = 4 \text{ to } 5; \frac{F_L}{F_I} = 3 \text{ to } 4.$$

This is plotted at A in Fig. 1.8.

(c) For the same LO and signal frequency ranges, as the IF changes to 20 MHz, the values of F_L/F_I and F_S/F_I each increase by a factor of 1.25. The area of interest moves through region C to area B in Fig. 1.8.

(d) From Fig. 1.8, the spurious responses (with $K \leq 5$) are
$$3F_L - 2F_S = F_I$$
at $F_L = 75$ MHz, $F_S = 100$ MHz, $F_I = 25$ MHz and at $F_L = 90$ MHz, $F_S = 125$ MHz, $F_I = 20$ MHz;
$$2F_S - 2F_L = F_I$$
at $F_L = 87.5$ MHz, $F_S = 100$ MHz, $F_I = 25$ MHz and at $F_L = 125$ MHz, $F_S = 135$ MHz, $F_I = 20$ MHz;
$$2F_S - 3F_L = F_I$$
at $F_L = 75$ MHz, $F_S = 125$ MHz, $F_I = 25$ MHz and at $F_L = 76\frac{2}{3}$ MHz, $F_S = 125$ MHz, $F_I = 20$ MHz;
$$2F_L - F_S = F_I$$
at $F_L = 75$ MHz, $F_S = 125$ MHz, $F_I = 25$ MHz.

(e) For the first three spurs above, the multiplier of F_S is 2, so a reduction in signal power of 10 dB will reduce the spur by 20 dB, or by 10 dB relative to the signal. For the last spur, the reduction will be 10 dB, so its ratio to the signal will not change. This assumes the signal is small enough, which may be verified by plotting the strength of the spur as the signal is varied over a range of powers.

The spur chart of Fig. 1.8 covers only a limited range of frequency ratios and may have to be expanded if the area of interest does not fall in this range. Also, higher-order spurs, especially those involving higher multiples of F_L, may be of importance. Note, also, that the main divisions of the graph (e.g., the vertical line at $F_S/F_I = 4$) correspond to solutions where n or m is zero. In a balanced mixer, the level of one input signal, usually the LO, is reduced due to the balance as it appears in the IF output. In a double-balanced mixer, both the LO and signal levels are reduced in the IF due to the balance. These reductions are also true of harmonics of the input signals.

1.4.2 Frequency Multipliers

Frequency multipliers are nonlinear devices which are optimized to produce certain desired harmonics of the input frequency.[8,9] These may be a single frequency or a whole "comb" of frequencies at every multiple of the input frequency within a certain frequency range. At lower frequencies, full wave rectifiers may be used as doublers and class-C amplifiers, tuned to a harmonic of the input frequency, can act as multipliers.[10] A nonlinear device which is commonly used in frequency multipliers is the step-recovery diode.[11,12] In general, a large number of harmonics are produced and filtering must be employed to reduce the undesired harmonics.

When a step-recovery diode is used and the output is concentrated at the Nth multiple of the input frequency, the power gain from input to multiplied output, neglecting circuit losses, has been estimated as

$$G_{\text{POWER}} \approx \frac{1}{N} \tag{1.17}$$

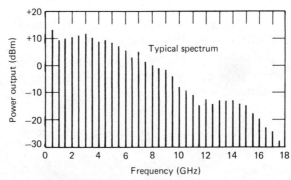

Figure 1.9 Spectrum of a comb generator. (From Hewlett-Packard, *Microwave* . . . Reprinted by permission of Hewlett-Packard Co.)

8 Kroupa, pp. 31–58.   **11** Hamilton.
9 Manassewitsch, pp. 342–350.   **12** "How to Select Varactors . . . ."
10 Kroupa, pp. 46–51.

for output frequencies above 5 GHz and about twice this value below 5 GHz.[13] If the output is untuned, the energy is spread over all multiples of the input frequency. A typical spectrum is shown in Fig. 1.9.

1.4.3 Frequency Dividers

There are a number of different types of frequncy dividers. Nonlinear circuits, which are similar to multipliers, may also be made to divide.[14-17] Oscillators may be synchronized at a subharmonic of the synchronizing frequency.[18-20] The scheme shown in Fig. 1.10 has also been used, although there can be problems in getting that circuit started.[21, 22] The most generally useful type of divider in frequency synthesizers is the digital divider.[23] These are composed of binary circuits which divide frequency by 2 but can be combined to divide by any whole number. Figure 1.11a illustrates the basic divide-by-2 and Fig. 1.11b shows a divide-by-4 circuit. With digital feedback, a divide-by-3 circuit can be created as shown in Fig. 1.11c. Chapter 6 is devoted to the study of frequency dividers.

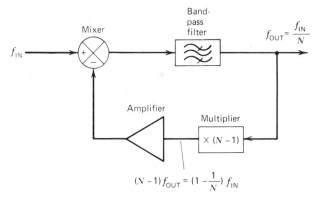

Figure 1.10   Regenerative divider.

PROBLEMS

1.1 The input signal to a radio receiver comes through a filter having a passband from 128 to 160 MHz. The local oscillator (LO) is higher than the signal to be received. (a) What range of LO frequencies is required to convert

13 Hewlett-Packard, from . . . .

14 Manassewitsch, pp. 361–370.

15 Penfield.

16 Miller.

17 Goldwasser.

18 Manassewitsch, pp. 353–355.

19 Kroupa, pp. 59–62.

20 Bearse.

21 Manassewitsch, pp. 352 and 353.

22 Kroupa, pp. 62 and 63.

23 Kasperkovitz.

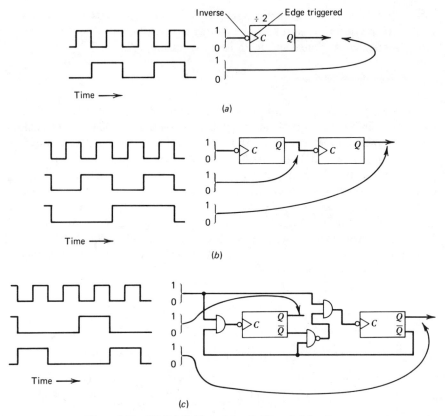

Figure 1.11   Digital dividers: (*a*) ÷2, (*b*) ÷4, (*c*) ÷3.

the input band to a 100-MHz intermediate frequency (IF)? (b) Plot the area of interest for this conversion on a chart like Fig. 1.8. (c) What spurious product occurs in this region? (d) If a 155-MHz signal is being received, what input frequency will produce a spur? (Hint: compute the LO frequency to convert the input signal to 100 MHz, then compute the input frequency to produce the spur with that LO frequency). (e) Plot the area of interest on a chart like Fig. 1.8 if the IF is extended from 90 to 110 MHz with no change in input signal and LO frequency ranges. (f) What additional spur(s) are now present in the area of interest?

1.2   A spurious signal occurs at the output of a mixer at a level of -50 dBm. Changing the signal by 1 kHz causes the spur to change by 4 kHz. (a) What multiple of the received signal is the spur (i.e., it is an $n \times m$ spur, where m represents the multiple of the LO. What is the value of n?). (b) If the signal amplitude decreases by 10 dB, what level would you expect the spur to assume? (c) What improvement in signal-to-spur ratio would you then expect?

1.3 Let

$$\alpha = \omega_1 t + \theta_1 \text{ and } \beta = \omega_2 t + \theta_2$$

and let

$$i_1 = I_1 \cos \alpha \text{ and } i_2 = I_2 \cos \beta. \tag{1.18}$$

If i_1 and i_2 flow in the same nonlinear resistance, the resulting voltage is

$$v = V_0 + ai + bi^2 + ci^3 + \cdots, \tag{1.19}$$

where

$$i = i_1 + i_2. \tag{1.20}$$

Using the trigonometric identity

$$\cos \alpha \cos \beta = \tfrac{1}{2} \left[\cos(\alpha + \beta) + \cos(\alpha - \beta) \right], \tag{1.21}$$

verify that expressions of the form of Eq. (1.8) are obtained by performing a complete expansion of the cubic term ci^3.

Suggested procedure:

(a) Substitute Eq. (1.20) into ci^3 from Eq. (1.19) and expand.

(b) Substitute Eq. (1.18) into the result.

(c) Use Eq. (1.21) and gather terms with similar powers of i_1 and i_2.

1.4 A phase detector operates at 10 kHz and 1 mV rms at 10 kHz appears at its output. The phase detector controls a VCO with a tuning sensitivity of 1 MHz/V. A low-pass filter is placed between the phase detector and the VCO to attenuate the 10-kHz signal. How much filter attenuation is needed at 10 kHz to produce a 20-Hz rms frequency deviation of the VCO output frequency at a 10-kHz modulation frequency?

2

Synthesizer Types

Now that some of the fundamentals have been considered, we begin the study of the synthesizer *per se* by describing the three most common types: the digital, or look-up-table synthesizer; the direct synthesizer; and the phase-locked, or indirect, synthesizer. The most commonly designed type is probably the latter and it is emphasized in this book. However, many of the techniques discussed are applicable to two, or to all three, types and it is important to understand the fundamentals of each, and some of its advantages and disadvantages, in order to be able to choose the appropriate type to use for a particular application.

2.1 THE DIGITAL (LOOK-UP-TABLE) SYNTHESIZER[1-3]

In this relatively new type of synthesizer, the waveform is synthesized piece by piece. A simplified block diagram is shown in Fig. 2.1. At the clock frequency, a number representing the phase change per clock period is shifted into the accumulator where it is added to the previous contents of the accumulator. The accumulator output is thus an approximation of a linear phase-versus-time function. It actually consists of equal discrete numerical changes at equal time intervals, as shown in Fig. 2.2 at A. The accumulator output is the address for a memory which contains numbers proportional to the cosines of its addresses. The capacity of the accumulator corresponds to one complete cycle. The memory output is converted to a voltage by the digital-to-analog converter (D-A) as illustrated for a system with very low resolution in Fig. 2.2 (curve D). If the capacity of the accumulator is

$$n_C = 2^{N_C},$$ (2.1)

---

1 Gorski-Popiel, pp. 39–44 and 121–149.
2 Fogarty.
3 Cooper.

14

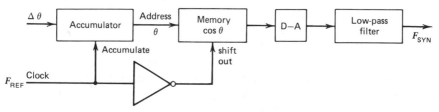

Figure 2.1 Basic diagram of a digital synthesizer.

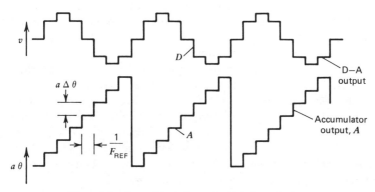

Figure 2.2 Some signals from the digital synthesizer.

and if n_i is added to the accumulator each clock interval, the number of clock intervals required to cycle the accumulator is n_C/n_i. Therefore, the rate at which the accumulator goes through a cycle is lower than the input frequency by that same factor and the synthesized frequency is

$$F_{\text{SYN}} = \frac{n_i}{n_C} F_{\text{REF}}. \tag{2.2}$$

The resolution is the minimum frequency change, which occurs when n_i changes by one:

$$\Delta f = \frac{F_{\text{REF}}}{n_C}. \tag{2.3}$$

EXAMPLE 2.1

Problem In a digital synthesizer, such as is shown in Fig. 2.1, the capacity of the accumulator is 2^{10}. What clock frequency is required to produce 1-kHz resolution (minimum step size)? Using that clock frequency, what number must be added to the accumulation at each clock time to produce a 37-kHz output?

Solution From Eq. (2.3)

$$F_{\text{REF}} = n_C \Delta f = 2^{10} \times 1 \text{ kHz} = 1.024 \text{ MHz}.$$

From Eq. (2.2),

$$n_i = \frac{F_{SYN}}{F_{REF}/n_C} = \frac{F_{SYN}}{\Delta f} = \frac{37\ \text{kHz}}{1\ \text{kHz}} = 37.$$

The highest practical synthesized frequency to allow reasonable low-pass filtering is about

$$F_{max} = \frac{F_{REF}}{4}. \tag{2.4}$$

Practical realizations can reduce the size of the memory, to cover only one quadrant of the cosine function, and still obtain the required output by proper mathematical manipulation of the address. Further memory reduction through increased mathematical complexity is also possible.

This type of synthesizer permits the phase to be set by setting the number in the accumulator. It can change frequencies very rapidly and fine resolution is relatively easy to attain. What is more, it is almost entirely digital. On the other hand, with currently available mass memories, the upper frequency is rather limited. This must be overcome by upconverting. To this end, two quadrature outputs may be provided to aid in achieving SSB upconversion. Spurious outputs due to imperfect D-A conversion can be a problem. The current limits of spurious suppression has been reported[4] as 50 to 60 dB.

2.2 DIRECT SYNTHESIZER[5-7]

The direct synthesizer employs multiplication, mixing, and division to generate a desired frequency from a single reference. The method is illustrated in Fig. 2.3.

A number of fixed frequencies are generated by multiplication, division, and mixing of a reference signal. Precision references are often at a frequency of 5 MHz because of desirable crystal properties in this range.[8] The reference might be divided to 1 MHz and then multiplied to 3 MHz and 27, . . . , 36 MHz, as required for Fig. 2.3. One of the frequencies from 27 to 36 MHz is switched into the first mixer where it is added to a fixed 3 MHz. The result is divided by 10 and, in the process, the variable digit is shifted from MHz value to 100 kHz value. The output of the divider is approximtely 3 MHz, so the scheme can be repeated indefinitely as illustrated in Fig. 2.3.

EXAMPLE 2.2

Problem   Show the switch closures and mathematical process to produce 6.154 379 MHz in the synthesizer illustrated in Fig. 2.3.

4   Gorski-Popiel, p. 44.   **7**   Manassewitsch, pp. 1–22 and 435–451.
5   Gorski-Popiel, pp. 26–32 and 47–68.   **8**   Manassewitsch, p. 485.
6   Kroupa, pp. 122–158.

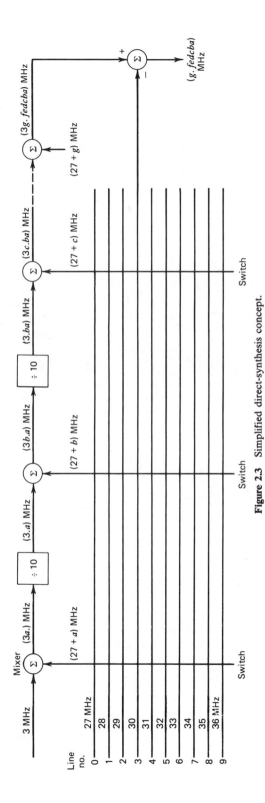

Figure 2.3 Simplified direct-synthesis concept.

17

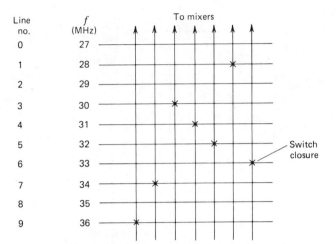

Figure 2.4   Switch closures for Example 2.2.

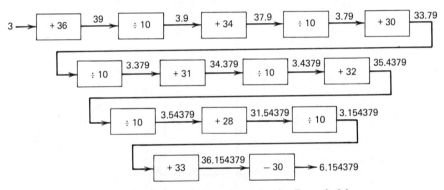

Figure 2.5   Mathematical representation for Example 2.2.

Solution   The switch closures are illustrated in Fig. 2.4 and the mathematical process is illustrated in Fig. 2.5.

The system illustrated in Fig. 2.3 is not completely practical because of the difficulty of removing the mixer input in the 27–36 MHz range from the output, whose range overlaps the input range. Figure 2.6 shows a practical mechanization, although all components (e.g., filters) have not been shown.[9, 10] These examples have been based on a decimal scheme, where one of 10 frequencies is chosen and the divider divides by 10. Other schemes (e.g., binary,[11] BCD[12]) have been used and a systematical method of design for various requirements has been developed.[13]

 9   Manassewitsch, pp. 442–451.   **12**   Papaiech.
10   Van Duzer.   **13**   Kroupa, pp. 122–150.
11   Oropeza.

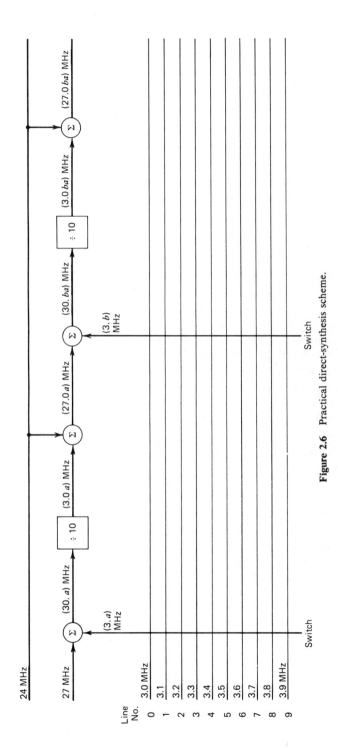

Figure 2.6 Practical direct-synthesis scheme.

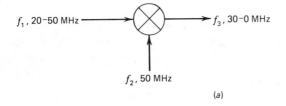

f_2, 50 MHz

(a)

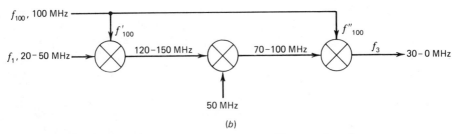

50 MHz

(b)

Figure 2.7 Two conversion schemes (filters not shown).

The direct synthesizer can change frequency rapidly, at the speed of the switches. Switching time can be increased by delays in the filters and by the necessity to filter the bias lines, but switching in a few microseconds is practical with this type of synthesizer.

The output spectrum close to the center can be very clean, essentaially a replica of the reference oscillator but with FM sidebands increased by the effective multiplication ratio from input to output.* Even if the output frequency is the result of many operations, its frequency is basically a fixed multiple of the reference, and FM sideband levels are as if the output was obtained by straight multiplication by the ratio of output to reference frequencies. It is possible, however, for varying delays in several paths to cause a different phase of the noise modulation to be effective at several points in the circuit.[14] To the degree that this occurs, additional noise can be added. For example, let us look at a circuit that might be used in a direct synthesizer but also has application elsewhere. Figure 2.7 shows two methods of achieving a frequency conversion. The first has a problem because the leaking input frequencies cannot be filtered from the output passband. The second method solves this problem. The output frequency can be written as

$$-f_3 = f_1 + f'_{100} - f_2 - f''_{100}, \tag{2.5}$$

$$f_3 = (f''_{100} - f'_{100}) + f_2 - f_1. \tag{2.6}$$

If $f'_{100} = f''_{100}$, the equation is the same as for Fig. 2.7a. To illustrate the problem that can occur, we write f'_{100} and f''_{100} with a single undesired modulation component as

$$f'_{100} = 100 \text{ MHz} + \Delta f \cos \omega_m (t + T')$$

---

* See Section 4.4 for a discussion of the effect of multiplication of FM sidebands.
14 Byers.

and

$$f''_{100} = 100 \text{ MHz} + \Delta f \cos \omega_m (t + T''),$$

where T' and T'' represent time delays from the source of modulation to the points of injection. The difference is then

$$f''_{100} - f'_{100} = \Delta f \left[\cos \omega_m (t + T'') - \cos \omega_n (t + T') \right] \tag{2.7}$$

$$= 2\Delta f \sin \omega_m \left(t + \frac{T' + T''}{2} \right) \sin \omega_m \left(\frac{T' - T''}{2} \right). \tag{2.8}$$

From this we see that imperfect cancellation can occur if the delays are not equal. This is more likely to be a problem at higher modulation frequencies.

Like the digital synthesizer, the direct synthesizer can have very fine resolution. From Fig. 2.6, its suitability for modularity is apparent, as is the ease with which the step size can be reduced. Commercial synthesizers have been made with specified nonharmonic spurs below -100 dB.* To achieve anything near this spur level, however, requires careful electrical and physical design.[15] The frequency scheme must be carefully chosen to avoid spurious products and the unit must be laid out with care to prevent cross coupling. Because of the large number of RF components and the required isolation, direct synthesizers tend to be bulky. They also seem to require considerable power, perhaps due to the large number of LO signals that must be produced.

2.3 PHASE-LOCKED (INDIRECT) SYNTHESIZERS

The phase-locked synthesizer, in effect, multiples the reference frequency by a variable number. It does so by dividing its output frequency by that variable number and adjusting the output frequency so that, after division, it is equal to the reference frequency. A simple loop is shown in Fig. 2.8a. Under conditions of lock, the two inputs to the phase detector have a constant phase relationship and must, therefore, have the same frequency. The output frequency F_{OUT} must therefore be

$$F_{\text{OUT}} = N F_{\text{REF}}. \tag{2.9}$$

If the output frequency increases, the frequency out of the divider ($\div N$) will exceed F_{REF}, the phase difference will drop and the phase-detector output will decrease. The tuning voltage to the VCO will decrease as a result and the output frequency will decrease, thus countering the frequency increase which began the sequence. The loop filter is present to suppress undesired components produced in the phase detector so they do not cause unacceptable FM on the VCO. The loop filter has an important effect on other noise, on acquisition of lock, on response speed, and on loop stability.

The mathematical control-system representation of the loop is shown in

---

*John Fluke Model 645A, Ailtech Model 360.
15 Gorski-Popiel, p. 31.

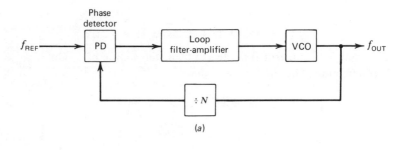

(a)

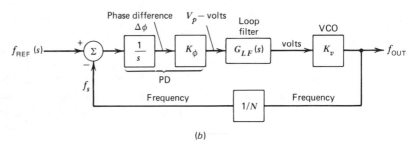

(b)

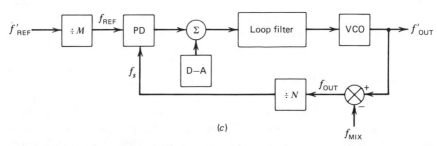

(c)

Figure 2.8 Synthesizer phase-locked loops: (a) basic synthesizer; (b) mathematical representation; (c) with added components.

Fig. 2.8b. The variable at the output is frequency. This is divided by N and subtracted from f_{REF}. The difference frequency is integrated, which is shown in Laplace notation as division by s, to give phase difference. The phase is converted to voltage in the phase detector. K_{φ} expresses the ratio of voltage to phase. The phase-detector-output voltage is multiplied by the loop-filter transfer function $G_{LF}(s)$, and the resulting voltage controls the frequency of the VCO. The ratio of frequency to control voltage is given by K_v. There is a one-for-one correspondence between the blocks in Figs. 2.8a and 2.8b except that the phase detector in Fig. 2.8a includes the frequency-subtracting junction and two blocks in Fig. 2.8b. Thus the voltage from the phase detector is proportional to the phase difference,

$$v_p = K_{\varphi}(\varphi_{REF} - \varphi_S), \tag{2.10}$$

which is related to the frequency difference:

$$v_p = K_\varphi (\int \int f_{\text{REF}} \, dt - \int \int f_s \, dt) \qquad (2.11)$$

$$= K_\varphi \int (f_{\text{REF}} - f_s) \, dt. \qquad (2.12)$$

Figure 2.8c shows some modifications of the block diagram which are not uncommon in practical synthesizers. None of these, in itself, changes the mathematical representation. The same basic loop could operate at a higher output frequency if the output were mixed down to the original frequency, thus leaving the values of N and f_{REF} unchanged. This frequency offset is not shown in the mathematical representation which is concerned only with response to a change from steady state and not with constant offsets which only affect steady state.

Often the reference frequency will be derived from a basic reference by division. This divider ($\div M$) usually need not be included in the mathematical representation.

A D-A converter may be included, as shown, to coarse tune the VCO to approximately the correct frequency in order to increase speed somewhat or to aid in acquiring lock. Once again, this is an additive signal and may not appear in the mathematical representation. However, if the response to a change in D-A output were being analyzed, this change would be shown as a step in voltage injected after the phase detector.

The phase-locked synthesizer is inherently slow compared to the direct and digital types. The frequency is changed by changing the divide number, N. This causes an error signal out of the phase detector and this in turn causes the frequency to change. Whereas, in the digital synthesizer, the frequency suddenly changes to a new value as the input to the accumulator ($\Delta\theta$ in Fig. 2.1) changes, and, likewise, in the direct synthesizer, there is a definite frequency change as the RF switches change the selected signals, the change in frequency of the phase-locked synthesizer is a continuous change, consisting of exponentials or damped sinusoids. The final frequency is never actually reached (except possibly momentarily) but only approached more closely as time goes on. The error signal changes value only once each reference period in most types and the loop bandwidth, which determines response speed, must be at most about one-tenth of the reference frequency. This is necessary for stability and to allow suppression of sidebands that are offset from spectral center by the reference frequency. The higher the reference frequency F_{REF} the faster may be the loop, but the speed of the phase-locked synthesizer can only approach that achievable with other types when its bandwidth is wide enough to make control of the phase shift in components, such as operational amplifiers, difficult, and only if the switching time is measured to a point where final frequency has not been too closely approached (i.e, a few time constants).

Spectral purity (the level of undesired sidebands) far from spectral center is basically that of the VCO and, sufficiently close to center, is that of the reference multiplied to the output frequency. Care must be taken to keep broadband noise in the loop components from corrupting the VCO spectrum. Grounding and shielding are of great importance here, as in the direct synthesizer, but here, rather than RF, the low frequencies that can modulate the VCO tuning voltage are most important. It is particularly difficult to suppress discrete sidebands separated from the output frequency by the reference frequency. Sidebands due to local low-frequency signals, such as ac power frequencies, are also difficult to suppress. While a few significant sidebands may appear relatively close to spectral center, the output of this synthesizer type will be relatively clean far from spectral center. First, the discrete sidebands are primarily at multiples of the reference frequency and tend to fall at least 12 dB/octave of separation from spectral center. Secondly, whereas the direct synthesizer output is basically that of a multiplied crystal oscillator and, therefore, reaches a plateau of noise at fairly small frequency offsets, the VCO noise of the phase-locked synthesizer continues to drop quite far from the spectral center and finally reaches a plateau which is relatively low both because of the higher power in the oscillator and because it is not multiplied.

Figure 2.9 shows the SSB density of synthesizers of both types. Also shown is the SSB level that would be obtained by multiplying the direct synthesizer output frequency to be equal to that of the indirect synthesizer (assuming all noise is FM). Note how high the noise floor would be. Also note how low the

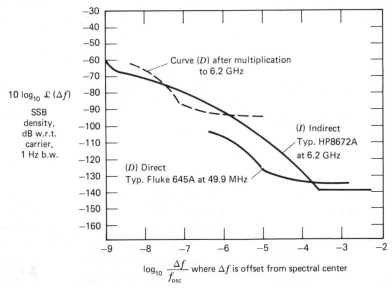

Figure 2.9 Typical SSB noise for two synthesizers (from manufacturers' data). $\Delta f / f_o$ is relative frequency offset.

noise of the phase-locked synthesizer can be made close to center. This probably entails the use of techniques described in the next section, plus great care in design.

The phase-locked synthesizer can often be made farily simple and is suitable for miniaturization and for low-power operation.

2.4 COMBINATIONS OF TECHNIQUES

Many synthesizers use combinations of the direct and phase-locked techniques. Combinations with the digital technique are also possible. For example, f_{mix} in Fig. 2.8c might be generated by direct synthesis and changed in a few large steps as the output frequency changes in many small steps over its range, so the frequency divider can cover a smaller range than the VCO. Such a technique is particularly applicable when the output is at microwave frequencies.

Often several loops are used and their outputs combined using techniques similar to those in a direct synthesizer. Figure 2.10 illustrates this concept. Here is a modular synthesizer, similar to that of Fig. 2.6, but employing phase-locked loops to generate the variable frequency for each decade. The speed and other advantages of a higher reference frequency are retained even though finer step sizes are generated. This is done by dividing the output frequency of one loop by a constant and mixing the result into another loop. Usually the divided output frequency must be converted before mixing with the next loop in order to permit proper filtering of the mixer products. Mixing inside the loop, as shown in Fig. 2.10, may provide improved suppression of mixer products compared to producing an output by directly mixing two synthesized signals. Also, a higher-level output signal is available from the VCO than from a mixer.

EXAMPLE 2.3

Problem In Fig. 2.10, list the divide numbers of the three loops (left to right) to produce 3.612 MHz. List the divide numbers to produce 3.123 MHz.

Solution The divide numbers for 3.612 MHz are 32, 48, and 53, as shown in Fig. 2.11. All frequencies are in megahertz.

From this, it can be surmised that the divide numbers for producing 3.abc MHz are, from left to right, $(30 + c)$ MHz, $(47 + b)$ MHz, $(47 + a)$ MHz. For 3.123 MHz, this implies divide numbers from left to right, of

$$33 \ 49 \ 48,$$

which can be proved by a diagram similar to Fig. 2.11.

More digits may be generated in one loop with a larger divide number at the output, but the mixing process may become more difficult because of the lower output frequency. This is illustrated in Fig. 2.12, where the same

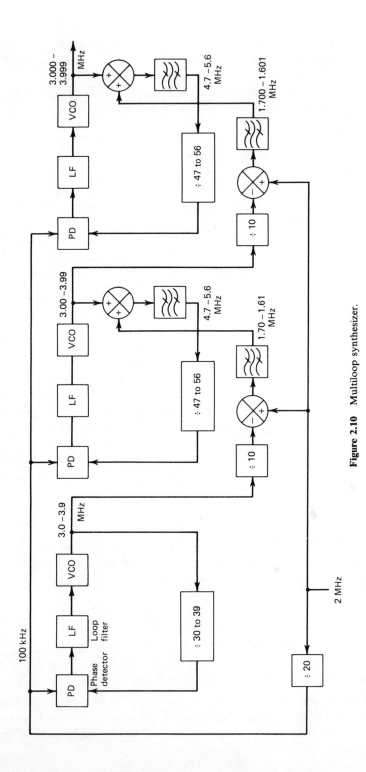

Figure 2.10 Multiloop synthesizer.

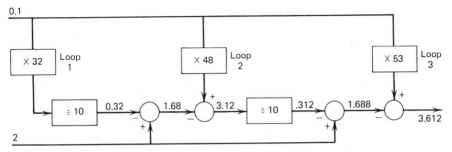

Figure 2.11 Mathematical representation for Example 2.3.

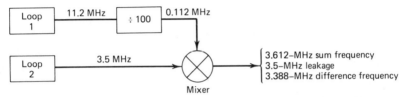

Figure 2.12 Combining two loops can create a difficult filtering problem. This one could probably be solved with a crystal filter if the output frequency of loop 2 were constant, but crystal filters are not available at much higher frequencies.

frequency that was produced by the three loops of Fig. 2.11 is produced by two loops. The difficulty involved in separating the desired sum frequency from the other two closely spaced frequencies at the mixer output is avoided in the system of Fig. 2.13. Here the output of a VCO at the desired sum frequency is mixed with the loop-2 output and the difference frequency is phase compared to the divided output from loop 1. The phase-detector output is used to control the VCO, thus forming a phase-locked loop. In this system,

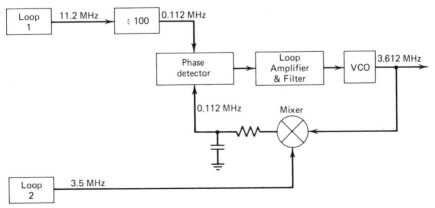

Figure 2.13 A method for combining two loops while avoiding filter difficulties. The VCO must be controlled to prevent it from locking 0.112 MHz below the loop-2 frequency.

the strong undesired mixer outputs are well separated in frequency from the desired signal, making the filtering problem much more manageable. Note that, regardless of the method used to combine the loops, as the output of loop 1 is divided by larger numbers, its required frequency range increases. If the loops are generating frequencies in 0.1-MHz steps, a loop whose output is divided by 10 must cover 1 MHz in order for the divider output to cover the 0.1 MHz between steps of the undivided loop. However, if a divide-by-100 is used, the loop must cover 10 MHz.

PROBLEMS

2.1 In a digital synthesizer, such as is shown in Fig. 2.1, the capacity of the accumulator is 2^8. What clock frequency is required to produce 10-kHz resolution (minimum step size)? Using that clock frequency, what number must be added to the accumulation at each clock time to produce a 300-kHz output?

2.2 In Fig. 2.3, beginning with the three switches shown, list the frequencies that must be selected to produce 137.1 kHz at the final output.

2.3 In Fig. 2.6, write the equations to show how 27.041 MHz is produced at the output, that is

$$3.1 \text{ MHz} + 27\text{MHz} = 30.1 \text{ MHz},$$
$$\frac{30.1 \text{ MHz}}{10} = 3.01 \text{ MHz},$$

etc. Then repeat these equations assuming that the reference has increased in frequency by 1% and show that the output frequency also increases by 1%.

2.4 In Fig. 2.10, list the divide numbers of the three loops (left to right) to produce 3.808 MHz.

3

Phase-Locked Synthesizer, Simple Loop, Linear Operation

We have considered the fundamental operations used in frequency synthesis and three different methods which use these operations to synthesize frequency. From this point, we will concentrate on the phase-locked method of synthesis. We begin by considering the operation of the simple phase-locked synthesizer in more detail, especially its linear operation, that is, how it responds to signals which are small enough that the loop parameters may be considered constant. Added complexities, which are important in practice, will be considered in subsequent chapters but, first, we need to establish a good understanding of the basics. The designer will often find himself returning to the consideration of the basic loop as a first step in a design or in determining feasibility.

3.1 ELEMENTARY DESCRIPTION OF COMPONENTS

We begin by describing, in some more detail than heretofore, the components used to implement the operations indicated by the blocks in the phase-locked synthesizer representations of Figs. 2.8a and 2.8b.

3.1.1 VCO[1–4]

The VCO may, in fact, be a current-controlled oscillator, an ICO. There is no basic difference insofar as the response of the loop is concerned. At low frequencies a current-controlled astable multivibrator, such as in Fig. 3.1, may be used.[5] The currents I_2 and I_3 control the charging rate of the capacitors and the frequency of the rectangular-output waveform is a linear function of these currents. These currents are in turn linearly related to I_1 and V_1 (although the relationship of frequency to V_1 is more temperature sensitive). This circuit has relatively poor spectral purity and has a potential start-up problem which may require additional circuitry to assure that both transistors do not remain saturated.

A second form of ICO, used at microwave frequencies, is the YIG-tuned oscillator.[6] Like the astable, this is inherently linear and can easily cover more than an octave frequency range. In addition, the YIG sphere has a very high Q.[7] High oscillator Q is necessary for good spectral purity and this requires high component Q. The disadvantages are that the magnetic circuitry which controls the resonant frequency of the oscillator is bulky and susceptible to stray magnetic fields. Isolation in the form of bulky magnetic shielding and physical separation from potential interfering fields may be necessary to

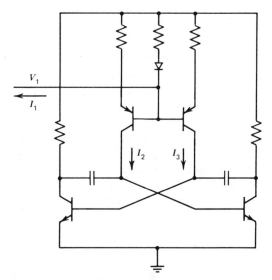

Figure 3.1 Current-controlled astable multivibrator.

1 Manassewitsch, pp. 370–398.	**5** Goodman.
2 Walston.	**6** Clark.
3 Kroupa, pp. 187–191.	**7** Tipon.
4 Solid state.	

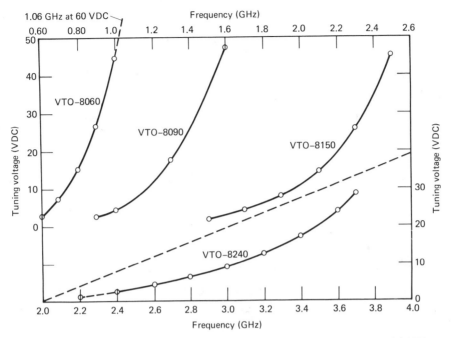

Figure 3.2 VCO tuning curves. (From Avantek Application Note ATP-1017, July 1974; reprinted with permission of Avantek, Inc., Santa Clara, CA.)

prevent frequency modulation of the oscillator. In addition, the frequency can be difficult to change rapidly, due to the inductance of the magnetic structure, and the tuning is characterized by some hysteresis.

Solid-state oscillators can be tuned by control of bias (supply) voltages which alter active-device capacitances or by changing bias on a varactor incorporated in the tuning circuit.[8,9] The varactor-tuned oscillator generally has a quite nonlinear tuning characteristic, especially when tuned over a wide frequency range. Figure 3.2 shows some typical tuning curves. The gain of the VCO block is

$$K_v \triangleq \frac{df_{out}}{dv_T} , \tag{3.1}$$

where f_{out} is the oscillator frequency and v_T is the tuning voltage. K_v is sometimes called the modulation sensitivity. This can easily vary by 10:1 over a usable tuning range. The loop gain will vary by the same amount due to this effect. This generally impedes optimization of loop performance. A linearizer

8 Herbert.
9 Solid state, p. 3.

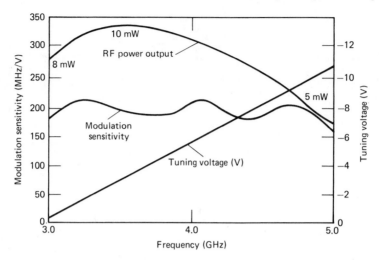

Figure 3.3 Tuning characteristics of a VCO using hyperabrupt varactor diodes. The tuning voltage is shown versus frequency as is the modulation sensitivity (K_v). Also shown is the output power versus frequency. (Reprinted by permission from Watkins-Johnson *Tech-notes*. See Beach, Fig. 6.)

may be used in series with the VCO to give a more constant combined transfer function. Linearizer design is discussed in Appendix 4A. Hyperabrupt junction varactors can give a much more linear tuning curve than ordinary varactors, as can be seen by comparing Fig. 3.2 with Fig. 3.3, but they tend to have lower Q.

To improve the linearity and noise performance of the VCO, its tuning range can be reduced if several VCO's are switched into operation, one at a time, or "fixed" tuning elements are switched into and out of the oscillator circuit.[10] The varactor can then be coupled to the tuned circuit less tightly and its losses will have less influence on circuit Q. It seems very likely that circuit Q is also reduced by dissipation of the RF signal due to forward biasing of the varactor, another condition improved by looser coupling. A reduction is also caused by some forms of limiting in the oscillator circuit.[11]

It has also been found that excessive impedance in the tuning line to the VCO causes an increase in noise. This may be due to leakage current across the varactor junction which becomes a source of noise if it is not "shorted-out" by a low impedance at the tuning source. For example, in Fig. 3.4, i_L is a noisy leakage current which will cause a voltage in series with the varactor equal to $i_L R_{\text{TUNE}}$.

Other methods of oscillator tuning also exist, particularly in microwave tubes: klystrons, magnetrons, and BWOs.

Where the required tuning range is only a few hundred parts per million or

10 Manassewitsch, pp. 381, 382, and 387–389.
11 Driscoll.

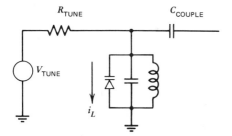

Figure 3.4 Varactor tuned circuit with leakage current.

less, a voltage-controlled crystal oscillator (VCXO) can be used.[12] This provides many of the advantages of the crystal oscillator, high Q and long-term frequency stability, although to a degree which is lessened by the voltage control. Voltage control is achieved by modifying the resonant frequency, which is primarily determined by the crystal, by varying the capacitance of a varactor which is also part of the resonant circuit.

Effects which tend to change the oscillator frequency, and thus lead to (open-loop) tuning inaccuracy and can introduce noise, are pushing, pulling, temperature sensitivity, and post-tuning drift. Pushing refers to the change in oscillator frequency with supply voltage. It is controlled by supply regulation and by supply filtering to reduce the introduction of noise. The pushing figure (e.g., in MHz/V) describes the magnitude of this effect. Pulling refers to the influence of the VSWR or load impedance on the tuned frequency. It is controlled by proper impedance matching. The pulling figure (e.g., in megahertz) gives the peak-to-peak frequency change as the phase of a given VSWR is varied over 360°. The resonant frequency is also a function of temperature. The term "post-tuning drift" describes changes in frequency which occur after a change in tuning voltage or current, as the oscillator is coming to equilibrium in its new state.[13] These are all open-loop characteristics and their effects are attenuated by closed-loop control, as in a synthesizer, but they must still be taken into account in design. For example, excessive open-loop frequency change due to these effects might prevent a loop from locking. In addition, in high-frequency oscillators, pulling can change the shape of the tuning curve and thus the loop gain.

3.1.2 Divide by $N(\div N)$

The frequency-divider ratio, N, is controlled by external logic signals which determine the value at which the digital divider is preset each cycle. The divider must be preset to a number which is higher for lower synthesized frequencies. This fact, plus the requirement for numerical offsets between the command word and the divide number, can lead to the use of logic to convert

12 Manassewitsch, pp. 389–398.
13 Buswell, "Voltage Controlled Oscillators . . . ."

from the available command word to that required by the divider. However, this can usually be avoided by techniques to be described later.

As can be seen from Fig. 2.8b, the ratio N enters directly into the loop gain. Thus, as the output frequency increases and N increases correspondingly, the loop gain, which is proportional to $1/N$, drops. Unfortuantely, this change is in the same direction as the usual change in VCO tuning slope with increasing frequency. These effects can be made to oppose each other by employing a frequency conversion between the VCO and divider which reverses the direction of frequency change. The forward gain of the analog circuitry can also be altered electronically to compensate for the gain change due to N.[14]

3.1.3 Phase Detector

In loops where the loop bandwidth can be much smaller than the reference frequency, a balanced mixer may be used for a phase detector. In most synthesizer designs, however, the loop bandwidth must be maximized in order to improve noise performance, speed, or acquisition range. This makes it difficult to filter undesired components which are produced in the phase detector at multiples of the reference frequency (which we can call reference sidebands), so it becomes very important to use a phase detector that produces little undesired energy. Most such phase detectors have a linear range of voltage versus phase covering 360° or 720°. An example of this type of characteristic is shown in Fig. 3.5. In this section we consider only operation along a single linear slope covering a 360° range. We defer to later consideration of the response to transients caused when the severe nonlinearities are transgressed.

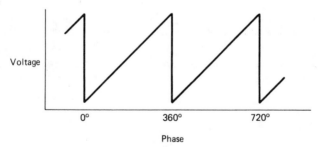

Figure 3.5 Typical phase-detector characteristic.

3.1.4 Loop Filter

The loop filter may vary from a simple amplifier, in cases where reference sidebands are not a problem, to a low-pass filter, to a circuit which includes an integrator. The most common types will be considered in Section 7.2. In

---

14 Egan, "LOs Share Circuitry . . . ," pp. 52 and 58.

this chapter we will concentrate on the simplest, the simple amplifier or the low pass, with or without an amplifier.

EXAMPLE 3.1

Problem A loop has a VCO with a tuning sensitivity (K_v) of 10 MHz/V. The loop filter is an R-C low-pass filter with a 1-kHz cutoff frequency. A balanced mixer is used for a phase detector with a reference frequency of 100 kHz. The reference frequency leaks through to the phase-detector output at a level of 0.01 V rms and the second harmonic of the reference appears in the output at a level of 1 V rms. What are the FM sideband frequencies and levels produced by these components at the synthesizer output?

Solution Due to the reference frequency, without the low-pass filter, the peak frequency deviation would be

$$\Delta f_{peak} = 0.01 \ V\sqrt{2} \ (10^7 \ Hz/V) = 10^5 \sqrt{2} \ Hz.$$

With the loop filter, this is reduced to

$$\Delta f_{peak} = 10^5 \sqrt{2} \left(\frac{10^3 Hz}{10^5 Hz} \right) = \sqrt{2} \ 10^3 \ Hz.$$

The corresponding modulation index is

$$m = \frac{\Delta f_{peak}}{f_m} = \frac{\sqrt{2} \ 10^3 \ Hz}{10^5 \ Hz} = \frac{\sqrt{2}}{100}.$$

This produces sidebands related to the carrier by

$$\frac{m}{2} = \frac{1}{\sqrt{2} \ 100} \rightarrow -43 \ dB \ at \ f_{out} \pm 100 \ kHz.$$

Due to the component at twice the reference frequency, the level of the sideband relative to the carrier is

$$\frac{m}{2} = \frac{\sqrt{2} \ V(10^7 \ Hz/V)}{2(2 \times 10^5 Hz)} \underbrace{\frac{10^3 \ Hz}{(2 \times 10^5 \ Hz)}}_{filter} = \frac{\sqrt{2}}{8} \rightarrow -15 \ dB$$

at $f_{out} \pm 200$ kHz.

3.2 LOOP OPERATION

Having described the components of the loop, let us now consider how they operate together to phase lock the VCO at a multiple of the reference frequency, first in qualitative terms and then mathematically.

3.2.1 Qualitative Description

Assume that the phase detector in Fig. 2.8*a* is of the sample-and-hold type, which changes output only at the completion of each output cycle of the divider, that is, each sample period. This will simplify the description of the loop's operation. Under steady-state conditions, the output of the $\div N$ circuit is at a frequency equal to f_{REF} so that a constant phase relationship is maintained between the two inputs to the phase detector. As a result, the phase-detector output is a steady voltage which is amplified in the loop filter and keeps the VCO tuned to a frequency NF_{REF}. Starting from this steady-state condition, let us consider how the loop responds to various disturbances.

One form of transient, which is of great interest, occurs when the value of N is changed to bring the synthesizer to a new frequency. If the value of N were increased, the first divider output transition affected by this change would occur later than it would have under the original steady-state conditions. This would cause a phase detector, with a characteristic such as shown in Fig. 3.5, to put out a higher voltage. If the loop filter consisted only of a frequency-independent gain, the VCO would immediately increase in frequency in response to the phase change. This would cause the time to the next sample, generated by a transition at the $\div N$ output, to shorten. If the gain about the loop were exactly right, the shortened period would exactly equal the period of F_{REF}. The next-sampled phase would then be the same as the previous value and a new steady-state would have been achieved in one sample period, as illustrated in Fig. 3.6b. Any lower gain would have increased the output frequency by a lesser amount, causing the phase-detector output to advance again at the end of the first period, as illustrated in Fig. 3.6c. The same process would then repeat itself each sample period, except that the changes would be smaller for each successive period. Thus would occur an exponentially shaped pull-in transient which would continually get closer to final value but would never quite make it. Conversely, a higher-than-optimum gain would cause the frequency correction to be excessive each time, so repeated overshoots would occur, as illustrated in Fig. 3.6d. If the gain were high enough, the oscillation would grow.

As the gain is reduced further and further, the response in Fig. 3.6c becomes slower and slower. Eventually, the change in phase-detector output from one sample to the next becomes small enough that the sampling process has little effect, and the speed of response is just a function of the loop gain.

In most practical synthesizers, the loop filter contains at least a low-pass filter. This would slow the transfer of the signal from the phase detector's output to the VCO and prevent the possibility of a perfect response such as described above for optimum gain. It also makes overshoot possible, even when the sampling rate is fast enough that its effects can be ignored. The overshoot occurs because the filter delays the loop's signal by an amount which is significant compared to the natural response speed of the loop, as determined by its gain. This has an effect similar to that described above, when the sampling rate is too low compared to the gain: the output frequency bypasses its final value before the influence of this new value of frequency can be felt to the degree necessary to prevent the overshoot.

A second way to increase the steady-state frequency would be to change the reference frequency. The response to a change in N is the same as the

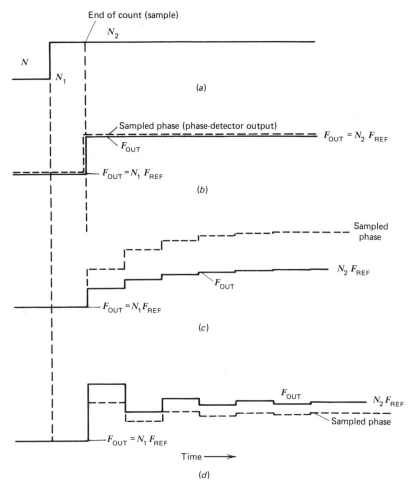

Figure 3.6 Responses of a simple loop: divide ratio at (a) and responses with (b) optimum gain, (c) low gain, and (d) high gain.

response to a step change in reference frequency. The new value of N is in effect duing the transient in either case. Since control theory is developed for changes in the reference, rather than in the loop-transfer functions, the equivalent change in reference input will be discussed in much of what follows.

A change in the reference frequency will eventually cause an N-times greater change in the output frequncy. Changes in the reference that occur slowly enough will be tracked almost exactly by the output. Changes which occur too fast will not: the response will fall at high modulation rates. However, the response can exceed the low-frequency response at certain frequencies that match the rate at which the loop tends to overshoot.

What would happen if the VCO frequency should suddenly change due to some outside influence, perhaps a supply-voltage transient? The same kind of reactions can be traced around the loop as in the case of a change in reference frequency. Here, however, rather than a delay occurring in the output-frequency change, there is a delay in the correction of that change. The outside influence causes the change, which is immediately seen at the output; then the loop responds to bring the output frequency back to its steady-state value. Thus, in this case, very low modulation frequencies are highly attenuated. The loop, which acts as a low-pass filter for variations in the reference, acts as a high-pass filter for variations in the VCO frequency.

Now that we have considered, in a conceptual way, how the loop responds, let us look at the same responses from a mathematical view.

3.2.2 Mathematical Description

The open-loop transfer function can be written with reference to Fig. 2.8b as the product of forward transfer function G and feedback H. In Laplace notation, we have

$$GH(s) = \frac{K_\varphi G_{\mathrm{LF}}(s) K_v}{Ns} .$$

(3.2)

The equivalent function of radian frequency is written

$$GH(\omega) = -j\frac{K_\varphi G_{\mathrm{LF}}(\omega) K_v}{N\omega} .$$

(3.3)

Let us assume initially that $G_{\mathrm{LF}}(\omega)$ is a constant equal to K_{LF}. Then Equation (3.2) becomes

$$GH(s) = \frac{K_F}{Ns} ,$$

(3.4)

where

$$K_F = K_\varphi K_{\mathrm{LF}} K_v.$$

(3.5)

K_F equals the change in oscillator frequency, in hertz, as a result of a change in phase of 1 cycle, for a linearized system (i.e., along a tangent to the tuning curve) at the operating point. It also equals the change of oscillator frequency in rad/sec for a 1-rad change in phase. It has units of seconds^{-1}. For example, if a phase change of 0.01 cycle causes a frequency change of 1 kHz (along a tangent to the tuning curve) then

$$K_F = \frac{10^3 \mathrm{Hz}}{0.01 \ \mathrm{cycle}} = 10^5 \ \mathrm{sec}^{-1}.$$

(3.6)

The control-system diagram is shown in Fig. 3.7a. The closed-loop response to

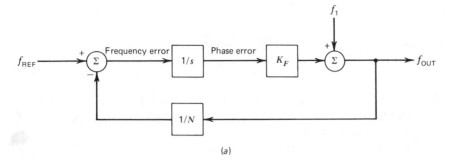

(a)

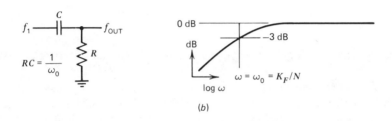

(b)

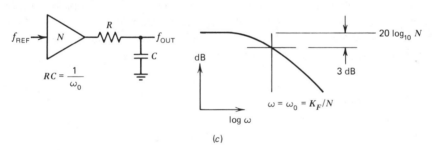

(c)

Figure 3.7 Simple loop (a) with equivalent circuits and frequency responses [(b) and (c)].

a signal at f_1 is

$$\frac{f_{\text{OUT}}}{f_1}(s) = \frac{G}{1 + GH} \qquad (3.7)$$

$$= \frac{1}{1 + K_F/(Ns)} \qquad (3.8)$$

$$= \frac{s}{s + \omega_0}, \qquad (3.9)$$

where

$$\omega_0 \triangleq \lim_{s \to 0} sGH(s) \tag{3.10}$$

$$= \frac{K_F}{N} . \tag{3.11}$$

This is equivalent to a high-pass filter, as shown in Fig. 3.7b. The 3-dB cutoff (the loop bandwidth) frequency is the frequency at which the open-loop gain has a magnitude of one and it equals ω_0.

The closed-loop response to a signal at f_{REF} is

$$\frac{f_{OUT}}{f_{REF}}(s) = \frac{K_F/s}{1 + K_F/(Ns)} \tag{3.12}$$

$$= N\frac{\omega_0}{s + \omega_0} . \tag{3.13}$$

This is equivalent to a gain of N plus a low-pass filter with cutoff at ω_0, as in Fig. 3.7c.

Thus a signal entering at f_{REF} will be amplified by N and passed to f_{OUT} with little attenuation at low frequencies but will fall off above the loop bandwidth, whereas a signal at f_1 will be attenuated below the loop bandwidth but pass to f_{OUT} with little attenuation above the loop bandwidth.

This simplest phase-locked loop is sometimes called a type-1 loop because the open-loop gain has one pole at zero frequency. It is also a first-order loop because the open-loop gain has one significant pole. If we add a low-pass filter, the loop will still be type-1, but it will become a second-order loop.[15] The loop-filter transfer function then becomes

$$G_{LF}(s) = \frac{K_{LF}}{1 + (s/\omega_p)} . \tag{3.14}$$

The open-loop transfer function is

$$GH(s) = \frac{K_F/N}{s[1 + (s/\omega_p)]} . \tag{3.15}$$

In terms of ω, it is

$$GH(\omega) = -j\frac{K_F/N}{\omega[1 + j(\omega/\omega_p)]} . \tag{3.16}$$

Figure 3.8 shows the control-system diagram for this loop, plus the corresponding Bode plot for phase and forward gain only. Also shown are the maximum and minimum values of N. Where these curves cross, the open-loop gain equals unity. In symbols

$$GH(\omega) = \frac{G(\omega)}{N} = 1 \tag{3.17}$$

15 Klapper, pp. 85 and 86.

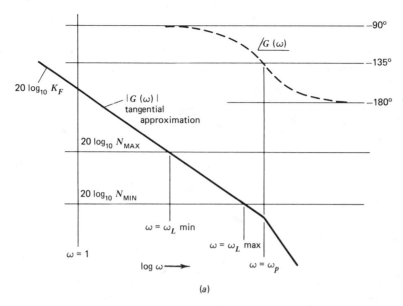

(a)

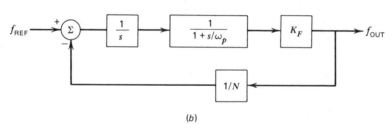

(b)

Figure 3.8 Simple loop (b) with bode plot (a).

implies

$$G(\omega) = N. \tag{3.18}$$

EXAMPLE 3.2

Problem Draw a Bode plot of the open-loop gain for a loop with the following parameters:

K_v, 10 MHz/V;
K_φ, 1 V/cycle;
Loop filter, R-C low-pass with 1-kHz cutoff frequency;
N, 1590.

Solution Below the filter corner frequency, the open-loop gain is

$$|GH(\omega)| = \frac{(10^7 \text{ Hz/V}) \, (1 \text{ V/cycle})}{1590 \, \omega} .$$

This becomes unity at

$$\omega = \frac{10^7}{1590} \ \sec^{-1} \ \text{or} \ f = \frac{\omega}{2\pi} = 1 \ \text{kHz}.$$

The Bode plot for $|GH(\omega)|$ is shown in Fig. 3.9.

The closed-loop transfer function from f_{REF} is

$$\frac{f_{\text{OUT}}}{f_{\text{REF}}}(s) = \frac{G(s)}{1 + GH(s)} \tag{3.19}$$

$$= \frac{NK_F}{Ns(1 + s/\omega_p) + K_F} \tag{3.20}$$

$$= \frac{\omega_p K_F}{s^2 + \omega_p s + \omega_p K_F / N}. \tag{3.21}$$

In terms of standard notation

$$\frac{f_{\text{OUT}}}{f_{\text{REF}}}(s) = \frac{\omega_p K_F}{s^2 + 2\zeta\omega_n s + \omega_n^2}, \tag{3.22}$$

where the natural frequency is

$$\omega_n = \left(\frac{\omega_p K_F}{N} \right)^{1/2} \tag{3.23}$$

$$= (\omega_0 \omega_p)^{1/2}, \tag{3.24}$$

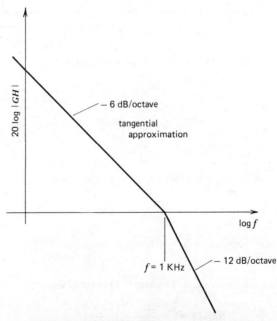

Figure 3.9 Bode plot of gain for Example 3.2.

the geometric mean of the loop bandwidth in the absence of a filter and the filter corner frequency. The damping factor is

$$\zeta = \tfrac{1}{2}\left(\omega_p \frac{N}{K_F}\right)^{1/2}$$

$$= \tfrac{1}{2}\left(\frac{\omega_p}{\omega_0}\right)^{1/2}. \tag{3.25}$$

The response from f_1 is

$$\frac{f_{\text{OUT}}}{f_1}(s) = \frac{s(s + \omega_p)}{s^2 + 2\zeta\omega_n s + \omega_n^2}. \tag{3.26}$$

The poles of the closed-loop function are located at

$$s_p = \zeta\omega_n \pm \omega_n(\zeta^2 - 1)^{1/2}. \tag{3.27}$$

These are shown in the root-locus plot of Fig. 3.10. As long as ζ is greater than one, the poles are real and a tangential plot of closed-loop gain looks as shown in Fig. 3.11. The characteristics are similar to the case with no loop filter (Fig. 3.7), except for the increasing rate of attenuation in $f_{\text{OUT}}(\omega)/f_{\text{REF}}(\omega)$ beyond approximately the filter corner frequency ω_p. As ω_p decreases toward ω_0, the damping ratio decreases and the phase shift at ω_0 increases. Correspondingly, the transient response of the loop becomes less damped (more ringing) and the frequency response peaks near ω_n, as can be seen in Fig. 3.12. From Fig. 3.13 it can be seen that, for large damping, the response is similar to that for no filter but, as the damping ratio decreases, the response peaks and the peak moves to a lower frequency relative to ω_0.

Relative stability of the loop is indicated by the phase margin, at unity open-loop gain, and the gain margin, where the excess phase is 180° (360° total around the loop). Figure 3.14 shows the phase margin as a function of damping factor. More highly damped loops are safer in that more parameter variation is allowable before instability occurs. From Fig. 3.8a or 3.10, we can see that, with a low-pass filter only, the loop is inherently stable, since a 180°

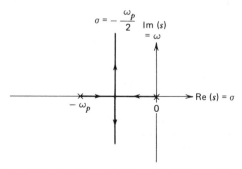

Figure 3.10   Root locus for loop with low-pass filter.

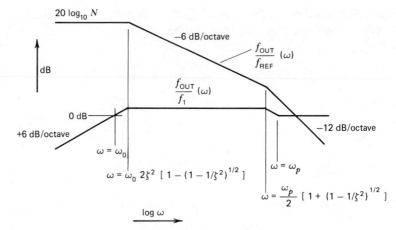

Figure 3.11 Tangential approximation for closed-loop gain, large ζ.

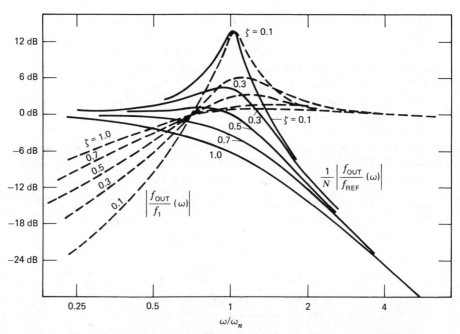

Figure 3.12 Closed-loop frequency response versus ω/ω_n, loop with low-pass filter.

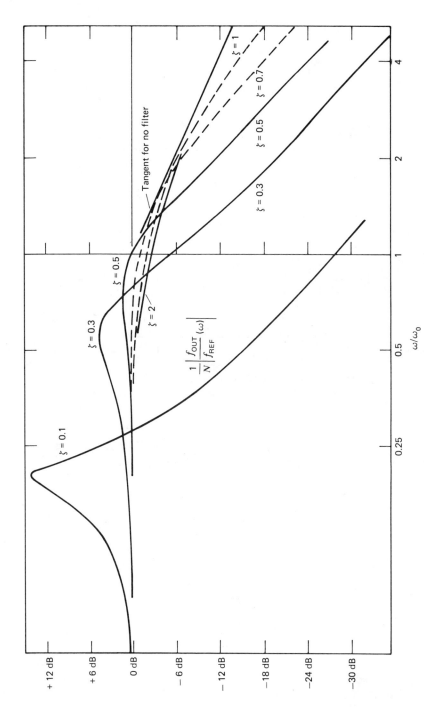

Figure 3.13 Closed-loop frequency response versus ω/ω_0, loop with low-pass filter.

$$\frac{1}{N}\left|\frac{f_{OUT}}{f_{REF}}(\omega)\right|$$

$\zeta = 0.1$

$\zeta = 0.3$

$\zeta = 0.5$

$\zeta = 2$

Tangent for no filter

$\zeta = 1$

$\zeta = 0.7$

$\zeta = 0.5$

$\zeta = 0.3$

ω/ω_0

+ 12 dB

+ 6 dB

0 dB

− 6 dB

− 12 dB

− 18 dB

− 24 dB

− 30 dB

0.25

0.5

1

2

4

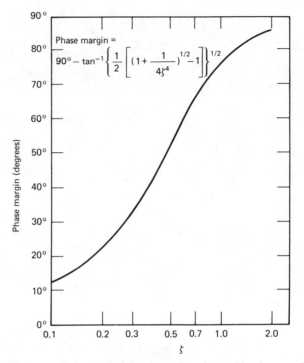

Figure 3.14 Phase margin versus damping factor for a loop with a low-pass loop filter.

phase shift cannot be attained for any finite frequency. However, it is an approximation to say that there is only a low-pass filter in a practical case because undesired poles inevitably exist at sufficiently high frequencies.

EXAMPLE 3.3

Problem For the loop of Example 3.2:

  (a) What is the damping factor?
  (b) What is the natural frequency?
  (c) Using the tangential approximation for gain, what is the phase margin?
  (d) If another pole is added to the loop filter at 10 kHz, what divide number will give zero phase margin? Use the Bode straight-line approximation for gain.

Solution (a) From Eq. (3.25):

$$\zeta = \tfrac{1}{2}\left(\frac{\omega_p}{\omega_0}\right)^{1/2} = 0.5.$$

  (b) From Eq. (3.24):

$$\omega_n = (\omega_0\omega_p)^{1/2}; \ f_n = \frac{\omega_n}{2\pi} = 1 \text{ kHz}$$

since $\omega_0 = \omega_p = 2\pi \ (1 \text{ kHz})$.

(c)   Phase margin is $90° - 45°$ (filter phase shift at corner) $= 45°$.

(d)   Phase shift due to both poles at a frequency f_x is

$$\varphi = \tan^{-1} \frac{f_x}{1 \text{ kHz}} + \tan^{-1} \frac{f_x}{10 \text{ kHz}} .$$

For zero phase margin this must be $90°$. By trial and error, or by plotting φ versus f_x, we find that zero phase margin occurs at 3.16 kHz:

$$\tan^{-1} \frac{3.16 \text{ kHz}}{1 \text{ kHz}} + \tan^{-1} \frac{3.16 \text{ kHz}}{10 \text{ kHz}} = 90°.$$

For unity loop gain at 3.16 kHz, we have

$$1 = \frac{K_\varphi K_v}{\omega N} \frac{\omega_p}{\omega} = \frac{10^7}{2\pi(3160)N} \frac{1000 \text{ Hz}}{3160 \text{ Hz}},$$

so $N = 159$. Alternately, referring to Fig. 3.9, to change frequency by a factor of 3.16 along a -12 dB/octave slope, requires a gain change of 3.16^2, or 10. This is accomplished by a reduction of divide ratio from 1590 to 159.

3.3   LOOP LOW-PASS FILTER

The purpose of the low-pass filter is to attentuate signals, at the reference frequency and its harmonics, which emanate from the phase detector and which cause FM sidebands on the VCO. In many designs, the loop filter is then constrained from being too high, because of the resulting poor attenuation at the reference frequency and its multiples, and from being too low, because of the resulting poor phase margin. The reference frequency and its lower harmonics may be of approximately equal amplitude but the lower-frequency components are usually more serious because they produce larger sidebands (i.e., the resulting frequency deviation may be the same but the modulation index is higher for the lower-frequency components), and they are attenuated less by a low-pass filter. A notch filter,[16, 17] which attenuates only a narrow band around an undesired frequency, will produce little phase shift far from its resonant frequency, but several of these may be required to suppress all of the undesired components. In some cases, the best solution may be a combination of one or more notch filters, to attenuate the reference frequency and its lowest harmonics, and a low-pass filter to reduce all of the components, particularly the higher frequencies, as well as broadband noise.

As poles are added to a low-pass filter, the corner frequency must be increased to maintain phase margin. Because of this, increasing the number of poles does not necessarily result in better filtering at a given frequency. Figure 3.15 shows the attenuation of various Butterworth and multiple-order-pole

16   Kroupa, pp. 220 and 221.
17   Manassewitsch, pp. 307–309.

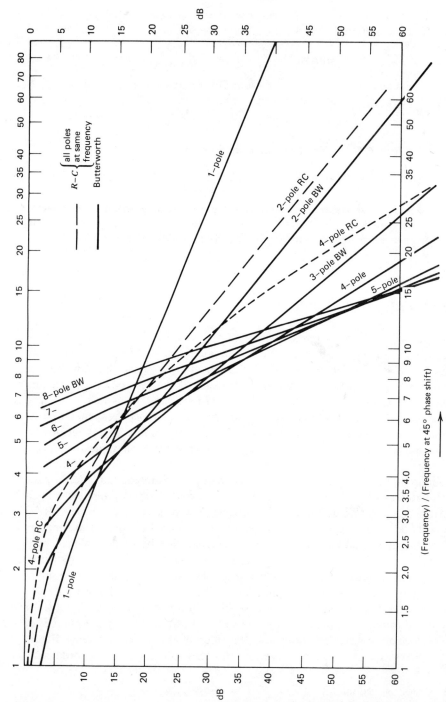

Figure 3.15 Low-pass filter attenuation versus offset from frequency at 45°.

filters. These are plotted as a function of frequency, normalized to the frequency that produces 45° phase shift. Thus, if 45° phase shift is allowed at a given frequency, these curves show what attenuation can be obtained with various filters. It can be seen that a single-pole filter is most effective at low frequencies. Actually, it is somewhat better than might be suggested by this normalized plot because it reduces the gain at the frequency giving 45° phase shift and this, in turn, would normally increase the phase margin at unity open-loop gain.

EXAMPLE 3.4

Problem If a signal at 10 kHz must be attenuated by a loop filter and no more than 45° phase shift is allowed to be produced by that filter at 2 kHz, how much attenuation can be obtained from a Butterworth filter and how many poles would it have?

Solution Check Fig. 3.15 at 10 kHz/2 kHz = 5. Here a three-pole filter gives 18 dB attenuation, but a two-pole filter is almost as effective.

3.4 HOLD-IN RANGE*

The hold-in range is the frequency change that would occur at the output of a locked loop if the phase-detector output were varied over its allowed range. If the elements following the phase detector are linear over its range, the hold-in range is related to the dc forward gain K_F by a constant which depends on the type of phase detector. Characteristics of typical phase detectors are shown in Fig. 3.16.

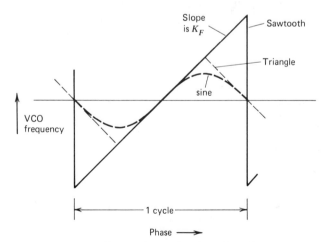

Figure 3.16 Characteristics of various phase detectors.

*This is the total, peak-to-peak, range. We will later describe this as $\pm\omega_H$ and ω_H is sometimes called the hold-in range, but ω_H is only one-half of the total discussed here.

For a sawtooth characteristic, the total peak-to-peak hold-in range equals K_F cycles:

$$\text{hold-in range} = K_F \text{ (1 cycle).} \qquad (3.28)$$

This is in hertz because K_F has units of seconds.$^{-1}$ A balanced mixer with square-wave inputs or an exclusive-OR gate produces a triangular characteristic. This is also approached when the inputs to a balanced mixer are equal in amplitude.[18] The triangular characteristic has half the hold-in range of the sawtooth for the same K_F, that is, $(K_F/2)$ cycle peak-to-peak.

A sinusoidal characteristic has a peak-to-peak hold-in range equal to $2K_F$ rad (where K_F is taken at maximum slope on the sinusoid, the normal design operating point), as can be seen from the following:

$$K_F = \left(\frac{d}{d\varphi_r} \frac{\text{hold-in range}}{2} \sin \varphi_r \right)\Bigg|_{\varphi=0}, \qquad (3.29)$$

where φ_r is in radians;

$$\text{hold-in-range} = 2K_F \text{ rad.} \qquad (3.30)$$

Thus, for the same forward gain, the sawtooth characteristic has π times more hold-in range than the sine characteristic and twice that of the triangle.

One means of increasing the tuning range over that permitted by the hold-in range is to add coarse tuning. This is done by adding a voltage to the phase-detector output, usually by means of a D-A converter, as shown in Fig. 2.8c. The D-A may be placed after the loop filter for increased tuning speed but noise originating in the D-A will not be filtered by the loop filter in that case. The D-A is controlled by the same digital tuning command that sets the divider ratio N. Individual control bits usually control switches within the D-A and the output is the sum of the individual switched signals.[19] The concept is illustrated in Fig. 3.17. The output is inherently a linear function of

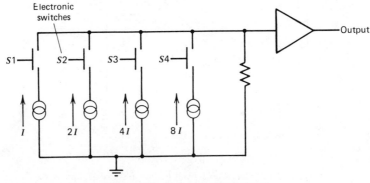

Figure 3.17   D-A converter concept.

18   Blanchard, pp. 8–29.
19   Manassewitsch, pp. 421–426.

the binary input and linearization of the VCO characteristic may be required for effective use of a D-A.

3.5 TRANSIENT RESPONSE

Since the most important transient is probably that caused by a change in divide ratio, we begin by verifying that the loop response to such a change is the same as its response to a step change in reference frequency.

When the divide number changes for N_1 to N_2, the divider output frequency changes from its original steady-state value, F_{OUT}/N_1, to F_{OUT}/N_2, which is larger by a factor of N_1/N_2. Since the original steady-state divider output had to equal F_{REF}, the divider output frequency just after the change must be $(N_1/N_2)\, F_{REF}$. Thus the frequency change at the divider output is $[(N_1/N_2) - 1]F_{REF}$ and the resulting change at the summing junction output is the negative of this, as illustrated in Fig. 3.18a. The same change can be produced at the summing junction output by a step of that size at f_{REF}, as shown in Fig. 3.18b. Since state varaibles are the same at the start

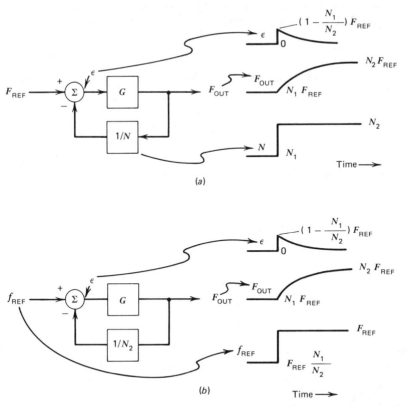

Figure 3.18 Identical outputs are produced by (a) a change in divide ratio and (b) a step change in reference frequency. (Note that F_{REF} is the actual reference frequency while f_{REF} is a change in the input to the loop.)

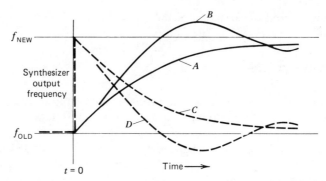

Figure 3.19 Transients when synthesizer is commanded to a new frequency.

of the transient for the two cases, and since the loop parameters are the same, the responses will be identical.*

The transient response of a loop without a filter, to a change in the value of N, or to a step in the value of f_{REF}, is the same as the response that would be obtained by driving the equivalent circuit of Fig. 3.7b with a step input. Such a response is illustrated in Fig. 3.19 (curve A). With a loop filter, overshoot and ringing can be produced, as shown in curve B.

EXAMPLE 3.5

Problem A frequency synthesizer with no loop filter has the following parameters:

$$K_F = 1 \text{ MHz/cycle},$$
$$F_{REF} = 10 \text{ kHz}.$$

It changes output frequency from 11 to 10 MHz. How long will it take from the time the divide number is changed until a frequency of 10.1 MHz is attained?

Solution Refer to the equivalent circuit of Fig. 3.7b. The time constant is $1/\omega_0$.

$$\omega_0 = \frac{K_F}{N} = \frac{10^6 \text{sec}^{-1}}{10 \text{ MHz}/10 \text{ kHz}} = 10^3 \text{ sec}^{-1},$$

$$\frac{\Delta f(T)}{\Delta f(0)} = e^{-\omega_0 T},$$

$$\frac{0.1 \text{ MHz}}{1 \text{ MHz}} = e^{-10^3 T \text{ sec}^{-1}}, \tag{3.31}$$

$$\ln \frac{0.1}{1} = -10^3 T \text{ sec}^{-1},$$

$$T = 2.3 \times 10^{-3} \text{ sec}.$$

---

*We assume that the divide ratio changes value such that it equals N_1 during one sample period (the period at the divider output) and N_2 during the next. The equivalent step in frequency must then occur at the beginning of the first period in which N_2 is effective. Actually, there is a delay involved in the effect of a step in reference frequency because the sampled phase depends on the reference phase at the *beginning* of the reference period, the start of the ramp in Fig. 3.5. A step in reference frequency would have to be applied at the beginning of the reference period during which the value of N changes, in order to have one full reference period during which to change the phase error (ϵ) by changing the time of the start of the next ramp. Our models will not take this delay into account, however, and it will usually not be significant.

Observe that the frequency error changes one decade each 2.3 time constants, that is,

$$T = \frac{2.3}{\omega_0} \log_{10} \frac{\Delta f(0)}{\Delta f(T)} \, . \tag{3.32}$$

The transient response of a loop without a filter to a step of frequency, introduced at f_1 in Fig. 3.7a, is the same as the response to an equivalent step input to a high-pass filter, as shown in Fig. 3.7b. The response is shown in Fig. 3.19 (curve C). Curve D shows a similar response for the underdamped loop that produced curve B.

By comparing curve A to curve C and curve B to curve D, we can see that the high-pass response and the low-pass response add together to give a step in frequency. In fact, we can show that, regardless of G_{LF}, the response to an excitation at f_{REF}, after division by N, when added to the response to the same excitation at f_1, equals the excitation. We show this as follows. The response due to $f(s)$ at f_{REF} is

$$f_{OUT, R} = \frac{f(s)G(s)}{1 + G(s)/N} \, . \tag{3.33}$$

The response to the same input at f_1 is

$$f_{OUT, 1} = \frac{f(s)}{1 + G(s)/N} \, . \tag{3.34}$$

Thus, the aforementioned combination gives

$$f_{OUT, 1} + \frac{f_{OUT,R}}{N} = f(s)\left(\frac{1 + G(s)/N}{1 + G(s)/N} \right) = f(s) \tag{3.35}$$

Because of this relationship, a plot of the transient response to a signal at f_1 also tells us the response to an input at f_{REF}, and vice versa. This relationship holds for both transient response and frequency response but is not so useful for the latter because it applies to the complex sum of the two responses, or to the sum of instantaneous values, and we are usually interested in the magnitude or envelope of the response.

If a D-A converter is incorporated after the phase detector, as shown in Fig. 2.8, a change in its output is equivalent to a step input at that point or an equivalent phase step prior to K_φ or an equivalent frequency impulse at f_{REF}, as illustrated in Fig. 3.20. For a loop with no filter, it is equivalent to a step at f_1. If the loop has no filter, or if the D-A output is inserted at f_1, the response to an accurate step from the D-A can add to the response due to a change in N to give an immediate change of f_{OUT} to its final value, regardless of loop gain.

Transient responses to various excitations, and for transfer functions corresponding to various loop filters, can, of course, be computed using Laplace transforms. The transient response of the general second-order loop will be discussed in Chapter 7. Tabulations of responses for various parameters are given there and elsewhere.[20, 21]

---

20 Blanchard, pp. 81–101.
21 Truxal, pp. 38–42.

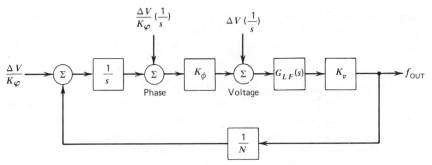

Figure 3.20   Equivalent Inputs for a D-A output of ΔV before the loop filter.

APPENDIX 3   LINEARIZER DESIGN[22-24]

We have seen (Fig. 3.2) that VCO tuning curves may be quite nonlinear. We have also noted that linearity may be desirable for effective use of a D-A and to allow better control of loop gain. With respect to the latter, consider that, at the point of maximum VCO sensitivity, the maximum forward gain might be limited by considerations of reference sideband level. Not only are these accentuated by high gain but the loop bandwidth increases due to high gain, forcing the loop low-pass filter corner to higher frequencies. With the maximum forward gain set by these considerations, the loop bandwidth can then become relatively narrow at some other frequency where the VCO sensitivity is low, leading to slow loop response and reduced suppression of VCO noise.

The purpose of the linearizer, which precedes the VCO, is to produce a linear net transfer function from linearizer input to VCO output.

Figure 3.21 shows a tuning curve for a VCO. The abscissa is VCO tuning voltage and the ordinate is VCO frequency. The desired linearizer characteristic is given by the same curve with the same abscissas but different ordinates. The abscissas are common because the VCO tuning voltage is the output of the linearizer block as well as the input to the following VCO block. The ordinate for the linearizer represents the linearizer input. If the same curve represents both transfer functions, then there is a linear relationship between the two ordinates, and that is what is desired. A scale factor or a constant offset may be applied to the linearizer input voltage without changing this representation.

For the linearizer characteristic to follow this smooth curve requires an infinite number of straight-line segments or a continuous function. We will consider two approximations, one using a finite number of straight lines and the second employing a continuous function.

The linearizer transfer function can be conveniently combined with that of the VCO in block diagrams used in analysis because the gain of the combination will be approximately constant.

22   Kincaid.
23   Buswell, "Linear VCO's."
24   Solid state $\cdots$, pp. 8–11.

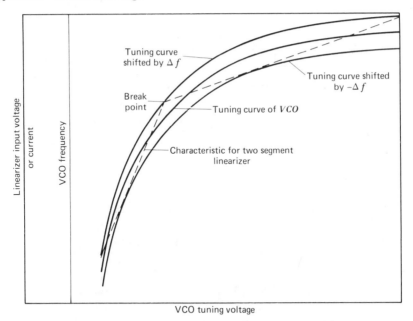

Figure 3.21   VCO and linearizer characteristics.

3A.1   Straight-line Approximation

Figure 3.21 shows a two-segment approximation. This is never further than a
constant offset, Δf, from the perfect tuning curve. Therefore, a given linearizer
input voltage will produce a tuning voltage that will produce a frequency that
is within Δf of the frequency of a linear characteristic.

Offsetting the tuning curve by $\pm\Delta f$ and choosing segments confined
between the offset curves, as in Fig. 3.21, provides a means for establishing a
linearizer curve which produces a given maximum frequency error. This is
appropriate to aid in accurate coarse tuning. If, however, we are more
concerned with constant gain, we can plot the log of the VCO tuning
sensitivity, as in Fig. 3.22, and divide the plot into equal segments such that
the change in $\log(df/dv)$ is the same for each segment. Then we have

$$\log_{10}\left(\frac{df}{dv}\right)_2 - \log_{10}\left(\frac{df}{dv}\right)_1 = K, \tag{3.36}$$

$$\frac{(df/dv)_2}{(df/dv)_1} = 10^K = \mathcal{K}, \tag{3.37}$$

where K and $\mathcal{K}$ are constant. Thus the linearizer can be designed to equalize
the geometric mean of the extremes of df/dv for each segment, and each
segment will then give the same ratio of actual to desired slope at its extremes.
The break points may also be obtained from this plot. A linearizer characteris-

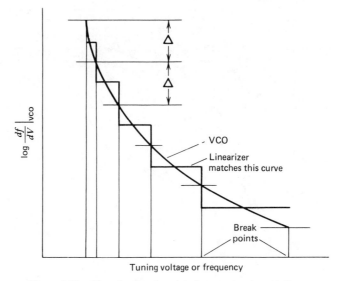

Figure 3.22   Choosing break points for equal gain variations.

tic so chosen may have to be modified some, however, to give good tuning accuracy in the sense of Fig. 3.21.

If constant loop gain is the main consideration, a shaper, rather than a linearizer, can be designed to distort the transfer function from shaper input to VCO output to compensate for gain changes due to the divider number N. To do this, a figure such as 3.22 can be drawn, but with the ordinate multiplied by $1/N$. The segments can then be chosen for equal variation in $(1/N)(df/dv)|_{VCO}$, and thus equal loop gain variations.

Figure 3.23 shows the concept of one type of linearizer. The input voltage is converted to current. At low levels of current, only the voltage divider with

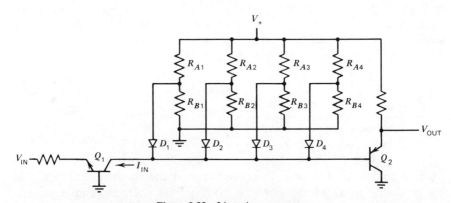

Figure 3.23   Linearizer concept.

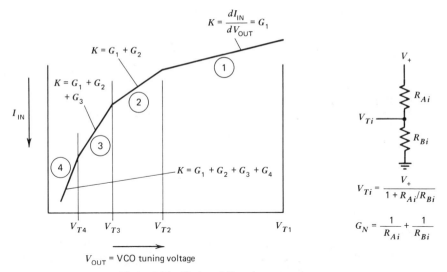

$$K = \frac{dI_{IN}}{dV_{OUT}} = G_1$$

$$K = G_1 + G_2$$

$$K = G_1 + G_2 + G_3$$

① ②

I_{IN} ③

④

$K = G_1 + G_2 + G_3 + G_4$

V_{T4} V_{T3} V_{T2} V_{T1}

V_{OUT} = VCO tuning voltage

$$V_{Ti} = \frac{V_+}{1 + R_{Ai}/R_{Bi}}$$

$$G_N = \frac{1}{R_{Ai}} + \frac{1}{R_{Bi}}$$

Figure 3.24 Choice of linearizer components.

the highest Thevenin-equivalent voltage conducts. The slope of the transfer function is established by the parallel resistance of these two resistors, say, R_{A1} and R_{B1}. The position of the segment depends on the Thevenin-equivalent voltage for the resistor pair. As the second-highest Thevenin voltage is reached, the corresponding diode conducts and the slope of the second segment is determined by the total conductance of the four resistors which are then connected to the conducting diodes. The process continues until all diodes are conducting. This is illustrated in Fig. 3.24. In designing the linearizer, one begins with the lowest-conductance slope, segment 1, and computes the open-circuit (Thevenin) voltage V_{T1} for $I_{IN} = 0$ and the conductance G_1 to produce the required slope. From these the values of the two resistors R_{A1} and R_{B1} are easily computed. The same process is used for subsequent segments. The voltage V_{Ti} is the break-point voltage at the low-current end of the ith segment where the diode D_i is just beginning to conduct. The required conductance is obtained by taking the difference between the slope of the segment under consideration and that of the previous segment. The values of the resistors for the ith segment can be easily shown to be

$$R_{Ai} = \frac{1}{G_i} \frac{V_+}{V_{Ti}}, \tag{3.38}$$

$$R_{Bi} = \frac{1}{G_i} \frac{1}{1 - V_{Ti}/V_+}. \tag{3.39}$$

The corners of the linearizer transfer function will be rounded somewhat by the nonlinear diode impedances. While the diode drops have not been considered, they are cancelled to a degree by the base-to-emitter drop of Q_2 in the configuration shown.

EXAMPLE 3A.1

Problem   In Fig. 3.24

$$V_{T3} = 12V,$$
$$V_+ = 40V$$
$$\frac{dI_{IN}}{dV} = 1 \text{ mmho (for segment 3)}$$
$$= 0.7 \text{ mmho (for segment 2).}$$

Find the values of R_{A3} and R_{B3}.

Solution

$$G_3 = 10^{-3} \text{ mho} - 0.7 \times 10^{-3} \text{ mho} = 0.3 \times 10^{-3} \text{ mho.}$$

Use Eq. (3.38) and (3.39):

$$R_{A3} = \frac{10^3}{0.3 \text{ mho}} \frac{40 \text{ V}}{12 \text{ V}} = 11.11 \text{ k}\Omega,$$

$$R_{B3} = \frac{10^3}{0.3 \text{ mho}} \frac{1}{1 - 12V/40V} = 4.76 \text{ k}\Omega.$$

3A.2   Continuous Approximation

Various nonlinear circuits can be used to approximate the required linearizer characteristic. In each case, care must be taken that the noise and the temperature sensitivity of the circuit are adequate for the application.

Figure 3.25 shows the tuning curve of one of the oscillators from Fig. 3.2 along with two approximations in which the tuning voltage equals the linearizer input voltage, raised to a power, plus a constant. Integrated circuits are now available which can produce a voltage proportional to some power of the input voltage, where that power may be selected over a restricted range.

The equations used to develop the two approximations shown in Fig. 3.25 follow. The approximation is, in general,

$$V = bf^\alpha + c, \tag{3.40}$$

where V is the tuning voltage; f is the frequency or the linearizer input voltage —recall that they are to be proportional; and b, c, and α are constants to be determined.

The derivative of V with respect to f at point 1, some point chosen on the tuning curve, is

$$V_1' = \alpha b f_1^{\alpha-1}. \tag{3.41}$$

Taking the ratio of two such values from different parts of the curve, we

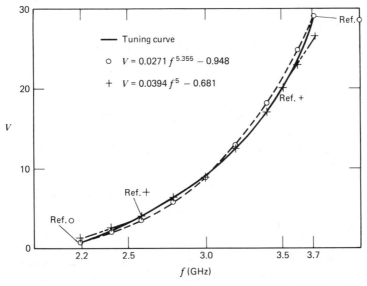

Figure 3.25 Two continuous approximations to a tuning curve.

obtain

$$\frac{V_1' f_1}{V_2' f_2} = \left(\frac{f_1}{f_2}\right)^\alpha, \tag{3.42}$$

and thus

$$\alpha = \frac{\ln\left[(V_1' f_1)/(V_2' f_2)\right]}{\ln(f_1/f_2)}. \tag{3.43}$$

After α has been computed, b can be obtained from

$$b = \frac{V_3 - V_4}{f_3^\alpha - f_4^\alpha}, \tag{3.44}$$

and c can then be obtained from

$$c = V_5 - b f_5^\alpha. \tag{3.45}$$

For the curves in Fig. 3.25, only two points were used to generate each curve, that is, V_1, V_3, and V_5 were taken at the same reference point, as were V_2 and V_4. These points are marked and it can be seen that the solutions are exact there.

PROBLEMS

3.1 Draw a Bode plot of the open-loop gain for a loop with the following parameters:

K_v, 5 MHz/V;
K_φ, 0.1 V/rad;
Loop filter, R-C low-pass with 1-kHz cutoff frequency;
N, 400.

3.2 For the loop of problem 3.1:
    (a) What is the damping factor?
    (b) What is the natural frequency?
    (c) Using the tangential (straight line) approximation for gain, what is the phase margin?

3.3 For the loop in problem 3.1, if another pole is added to the loop filter at 10 kHz, what divide number will give zero phase margin? Use the tangential (straight-line) approximation for gain.

3.4 In the loop of Problem 3.1, a sample-and-hold phase detector is employed with a reference frequency of 50 kHz. The reference frequency leaks through to the phase-detector output at a level of 1 mV rms. What are the FM-sideband frequencies and levels produced by this component at the synthesizer output?

3.5 If a signal at 8 kHz must be attenuated by a loop filter and no more than 45° phase shift is allowed to be produced by that filter at 1 kHz, what is the maximum attainable attenuation from a Butterworth filter and how many poles would it have? How much poorer is a filter with the same number of poles, all at the same frequency, under these same constraints?

3.6 Derive Eq. (3.32).

3.7 A frequency syntheizer with no loop filter has the following parameters:

$$K_F = 2\pi 10^4 \ \sec^{-1},$$

$$F_{\text{REF}} = 1\text{kHz}.$$

It changes output frequency from 90 to 100 kHz. (a) How long will it take from the time the divide number is changed until a frequency of 99.90 kHz is attained? (b) How much total time would be required to reach 99.99 kHz?

3.8 Repeat Problem 3.7 with K_F doubled.

4

Modulation, Sidebands, and Noise Spectrums

The output of a frequency synthesizer, ideally, is a single signal, constant in both amplitude and frequency. Thus amplitude and phase modulation and spurious signals are contaminants which must be controlled. The designer must understand not only the sources of these contaminants, but how they are affected in passing through various types of circuit elements, in order that he can both control the purity of the synthesizer's output during the design process and determine the degree of purity that is required for a particular application.

The detailed requirements for spectral purity should be established before the synthesizer design begins, otherwise the design effort may be wasted on a synthesizer which will not meet the system requirements. Sometimes those establishing the synthesizer requirement may not be aware of the kind of contaminants that are likely at the synthesizer output, so these may not be properly covered by specification. At other times, the synthesizer designer may be best qualified to establish these specifications because of his understanding of the effect of these contaminating signals upon the system. There is an advantage in the synthesizer designer understanding the use to which the synthesizer will be put because he may then be able to arrange that the specifications not be unnecessarily restrictive. For example, he may be able to avoid the use of certain frequencies, within the synthesizer, to which the system would be particularly sensitive and to obtain a reduction in specifications in the range of frequencies that he does use, thus easing his design task.

Once the requirements for spectral purity have been established, the designer will refer to them repeatedly in determining the architecture of the synthesizer, the types of circuits to be used and the detailed requirements on the circuits and even the components within those circuits. A change in the specified spectral purity of a synthesizer can easily double its complexity.

The sources of contaminating noise within the synthesizer will be considered, with particular emphasis on the noise spectrums of the oscillators. Ultimately, it is these oscillator spectrums which are combined to produce the synthesizer's output spectrum.

The manner in which contaminating signals are acted upon by the various types of circuit elements will also be studied. The suppression of the coupling of undesired signals by stray paths will be discussed in Chapter 10, but here we will be concerned with undesired signals and noise which accompany the desired signal along its intended path. Although contaminating, or undesired, signals are, by definition, noise, we will use "contaminating signals" to suggest undesired energy which is characterized by discrete frequencies. Discussions of the processing of such signals, however, also apply to the processing of the components of broadband noise at these same frequencies.

4.1 SPECTRAL REPRESENTATION OF AM AND NARROW-BAND FM

An amplitude-modulated (AM) signal is illustrated in Fig. 4.1a and its Fourier spectrum is shown in Fig. 4.1b. A third representation, by phasors, is shown in Fig. 4.1c. Here, the unit phasor is rotating counterclockwise in the complex plane and its real part represents the signal. In effect, the plane in which the phasor is represented may be viewed as rotating clockwise at the carrier frequency f_C, so the phasor is viewed as stationary. At the tip of the phasor are two other phasors which represent the sidebands at $f_C \pm f_m$ and which are therefore rotating with respect to the unit phasor at the modulation frequency f_m. The magnitude of the signal is equal to the magnitude of the sum of these three components and the signal is equal to the projection of this sum vector on a line which is rotating clockwise at the carrier frequency. The phasor diagram represents amplitude modulation because the two smaller phasors always add to give a resultant in the same direction as the unit phasor and the sum, therefore, is only changed in magnitude, not in phase.

A similar representation for FM is shown in Fig. 4.2. The magnitude of the Fourier component at the carrier frequency is proportional to the peak amplitude A and to the zero-order Bessel function $J_0(m)$ where m is the peak phase deviation in radians, called the modulation index. Its relationship to the peak frequency deviation Δf is obtained from

$$\varphi_r = \int \omega \, dt. \qquad (4.1)$$

The subscript r will be used throughout this book to indicate phase in radians, whereas the subscript c will be used to indicate phase in cycles. For a

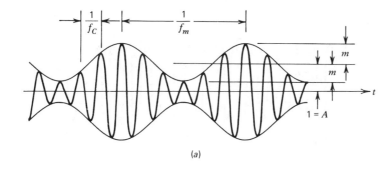

(a)

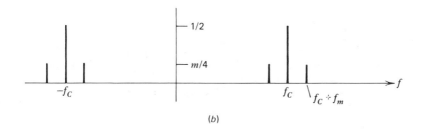

(b)

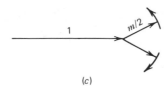

(c)

Figure 4.1 AM: (a) time, (b) Fourier, (c) phasor representations.

sinusoidal frequency modulation, the instantaneous frequency deviation

$$f = \Delta f \cos(2\pi f_m t + \theta) \tag{4.2}$$

implies a phase deviation,

$$\varphi_r = \int 2\pi \Delta f \cos(2\pi f_m t + \theta)\, dt \tag{4.3}$$

$$= \left(\frac{\Delta f}{f_m}\right) \sin(\omega t + \theta). \tag{4.4}$$

Thus FM, with peak frequency deviation Δf and modulation frequency f_m, implies PM (phase modulation) at the same modulating frequency and with a peak phase deviation of

$$m = \frac{\Delta f}{f_m}. \tag{4.5}$$

The magnitude of the nth FM sideband is $AJ_n(m)$, where n is zero for the carrier. For small values of m, we approximate the Bessel function magnitudes

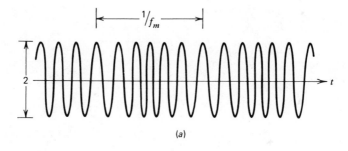

(a)

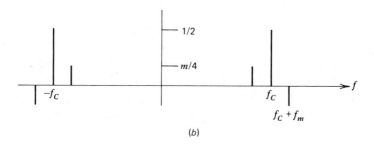

(b)

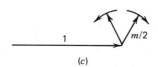

(c)

Figure 4.2 Narrow-band FM (exaggerated for illustration): (a) time, (b) Fourier, (c) phasor representations.

as

$$J_0(m) \approx 1, \tag{4.6}$$

$$J_1(m) \approx \frac{m}{2}, \tag{4.7}$$

$$J_n(m) \approx 0 \text{ for } n \geq 2. \tag{4.8}$$

This relationship between the peak phase deviation and the sideband level can be seen from Fig. 4.2c. For a modulation index of 0.1, the error in the approximation is 0.25% for $J_0(0.1)$, and 0.12% and $J_1(0.1)$ and $J_2(0.1)$ is about 0.1% of $J_0(0.1)$ and 2.4% of $J_1(0.1)$.

EXAMPLE 4.1

Problem   Refer to Fig. 1.6. Plot the phase of the signal at point D and the voltage at point A on the same time scale.

Solution   The magnitude of the signal at point A is given in Example 1.1 and illustrated in Fig. 4.3. Using the transfer function from A to D and Eq (4.1), the phase deviation at point D is plotted in Fig. 4.3.

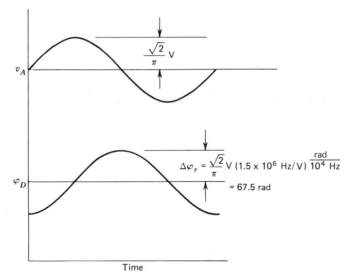

Figure 4.3 Waveforms in Example 4.1.

4.2 DECOMPOSITION OF SSB INTO AM AND FM

Because, in many cases, the response of a device to AM and to FM may be known better than its response to two simultaneous signals, it is useful to be able to decompose a larger signal plus a smaller single sideband (SSB) into AM plus FM on a carrier.

Figure 4.4b shows that a small SSB signal can be looked upon as a combination of AM and FM on the carrier.[1,2] In Fig. 4.4c the same thing is demonstrated through a phasor representation. As can be seen from both representations, the modulation sidebands on one side add while those on the other side are cancelled. Therefore, we can look upon a small single sideband as a combination of AM and FM, both the AM and FM sidebands having one-half the amplitude of the single sideband.

Note that when we decompose SSB into AM and FM (or PM), there is still only one carrier and its level does not change. There is not a carrier for AM and a separate carrier for FM. Consider the following. Start with a carrier. Amplitude modulate it with small m. AM sidebands are created, as shown above, but the carrier level does not change. Then apply frequency modulation with the same value of m. Since m is small, the carrier is not reduced appreciably ($J_0 \approx 1$) and FM sidebands are created at the same frequencies as the AM sidebands. FM sidebands are also created about the AM sidebands, but their amplitudes are $\frac{1}{2} m$ relative to the AM sidebands, whose amplitudes

1   Manassewitsch, pp. 82–86.
2   Goldman.

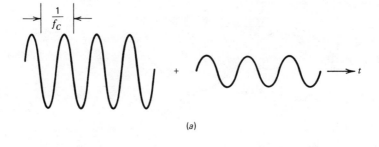

(a)

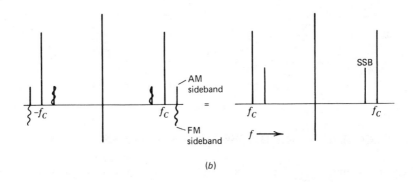

(b)

(c)

Figure 4.4 Decomposition of SSB into AM and FM: (*a*) time, (*b*) Fourier, (*c*) phasor representations.

are $\frac{1}{2} m$ relative to the carrier, so we ignore these. Therefore, we have AM and FM sidebands on the same carrier and its amplitude is essentially the same as without the sidebands.

We should be aware that, in some devices, AM can cause PM and, where a device's passband is not flat, FM will produce AM. These phenomena can, of course, adversely affect the correspondence between observed performance and any analysis that does not take them into account.

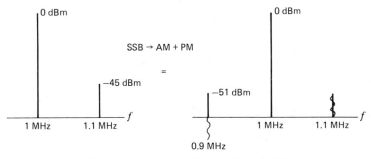

Figure 4.5 Decomposition for Example 4.2.

EXAMPLE 4.2

Problem A 0-dBm signal at 1 MHz and a −45-dBm signal at 1.1 MHz enter a limiting device which removes all amplitude modulation. The output contains a −10-dBm signal at 1 MHz. What other component(s) should we expect to find and at what amplitude(s)?

Solution See Fig. 4.5. The limiting device is said to remove AM, so we eliminate the AM sidebands. We then have only FM sidebands on a 1-MHz carrier. Since the carrier is reduced 10 dB in amplitude by the limiting device, we expect to find the FM sidebands reduced 10 dB. Otherwise we would have to assume some filtering and none was specified. That is, attenuating an FM signal changes neither its frequency deviation nor its modulation frequency, so its relative sideband level (−51 dB) will not change. Therefore, we expect to find, at the limiter output, the levels given in Fig. 4.6.

Some limiters are basically nonlinear devices, which produce harmonic and other components. One form of "limiting device" is an automatic gain control circuit, which adjusts an amplifier gain to keep the output level constant. If the circuit is fast enough compared to the modulation rate, this circuit can eliminate AM without producing significant harmonics or other products. All we were told about the limiting device in this problem is that it removes AM. Therefore, the solution is based on this simple assumption.

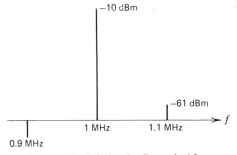

Figure 4.6 Solution for Example 4.2.

4.3 EFFECT OF MODULATION OF THE LOCAL OSCILLATOR IN A MIXER

A mixer normally acts in a basically linear manner with respect to the amplitudes of its signal components. Such is not the case, however, with respect to the local oscillator. The LO is supposed to be a simple single-frequency signal in almost all applications. However, it is important to understand the effects upon the IF of contaminating signals which often exist along with the desired LO.

4.3.1 FM Transfer

Since the mixer produces an IF whose frequency equals the sum or difference of the LO and the signal frequencies, FM is transferred from the LO to the IF, as illustrated in Fig. 4.7. FM sidebands are thus also transferred from LO to IF. FM sidebands on the LO which are 30 dB below the carrier produce sidebands on the IF which are 30 dB below the carrier, as shown in Fig. 4.8.

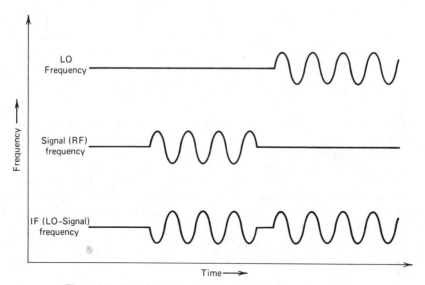

Figure 4.7   Transfer of frequency modulation through a mixer.

4.3.2. AM Suppression

Amplitude modulation is also transferred from LO to IF but the relative level of the sidebands is generally reduced. The reason is that, in order to reduce insertion loss, the mixer is usually operated with sufficient LO power to place

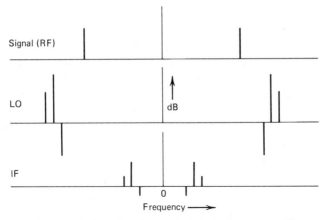

Figure 4.8   Transfer of an FM spectrum from LO to IF.

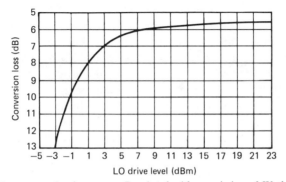

Figure 4.9   A mixer conversion-loss curve. Reprinted with permission of Watkins-Johnson Co.

the operating point on the flat part of the conversion-loss–versus–LO-power curve shown in Fig. 4.9. Thus, changes in the strength of the LO cause relatively small changes in the strength of the IF signal.

4.3.3   Mixing Between LO Components

In a single-diode mixer, the LO and signal are added, so it makes no difference whether a signal comes from the LO or signal input. In a mixer in which the LO is balanced, however, mixing products between the LO and other signals which enter the LO port are also balanced, and, therefore, produce IF components which are weaker than if the other signals had entered the signal (RF) port. The relative reduction depends on the degree of balance, but typically may be about 20 dB. When several signals enter the signal port, they also mix but their products tend to be weaker because none is usually as strong as an LO.

FM sidebands on the LO mix with the LO to produce IF signals which cancel each other (see Fig. 4.2b). This makes sense when we consider the LO and its sidebands as a single signal with changing instantaneous frequency. We would not expect a single signal to produce mixing products (except harmonics). A pair of AM sidebands, on the other hand, will mix with the LO to produce signals which add (this is essentially a detection process).

4.3.4 Contamination of the IF

Undesired sidebands on the LO can be troublesome if they are separated from the LO by less than the average of the RF (signal input) and IF (output) bandwidths because such modulation components which are transferred to the IF will pass through the IF filter. Such sidebands are usually a more serious problem if they are FM than if they are AM, not only because AM can be partially balanced out, but because, as will be seen, FM is generally stronger on an oscillator or synthesizer output than AM. These sidebands in the IF may be considered "single-frequency spurs." They occur if, and only if, a signal is present in the mixer to produce an IF to which they can be transferred, and their level is proportional to the level of that signal.

The most troublesome undesired signal can be an AM (or SSB) signal at the LO port, separated from the LO frequency by the IF. Even though the resulting mixing product may be partially balanced, it tends to be large because it results from mixing with a powerful LO. Such a signal is independent of the signal input to the mixer and is sometimes called an "internal spur".

These types of spurious products are illustrated in Fig. 4.10.

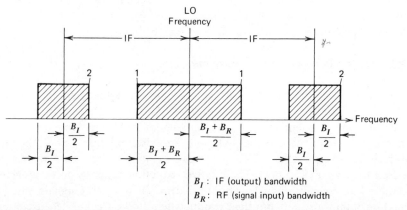

Figure 4.10 Regions of troublesome sidebands on the LO. Region 1 produces sidebands on the IF signal which can fall in the IF passband (even if the IF signal does not). Region 2 mixes with the LO to produce a signal in the IF passband.

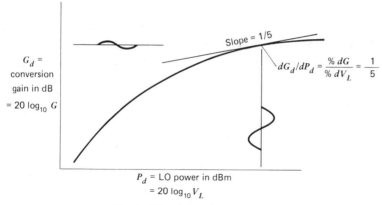

Figure 4.11 Conversion loss curve.

EXAMPLE 4.3

Problem   A 100-MHz LO enters a mixer at $+10$ dBm along with a 110-MHz single spurious sideband at -20 dBm. The mixer conversion gain curve has a slope of $\frac{1}{5}$ (see Fig. 4.11) at the operating point. (1) How much AM (in percent modulation) appears on the 10-MHz mixer output? (2) How much peak FM frequency deviation? (3) If the mixer has 10-dB conversion loss and -20-dB unbalance for LO-port signals, what signal (magnitude and frequency) would be expected below 15 MHz at the mixer output?

Solution   The decomposition of the single sideband into AM and FM is shown in Fig. 4.12.

  (a)   AM on LO:

$$20\log_{10}\frac{m}{2} = -36 \text{ dB},$$

$$m = 2 \times 10^{-36/20} = 0.032 \to 3.2\%.$$

AM on the output is caused by changes in conversion gain (the ratio of the IF output strength to the input signal strength) as the LO level changes. The change in conversion gain is given as $\frac{1}{5}$ the change in LO strength. The slope

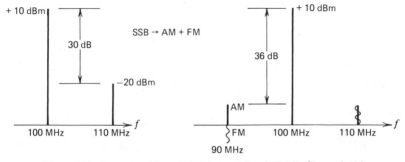

Figure 4.12 Decomposition of SSB into AM and FM for Example 4.3.

of a log-log curve (dB vs. dB) equals the ratio of percent changes at that point. Thus a small percentage increase in LO voltage causes the IF output to increase by $\frac{1}{5}$ of that percentage, if the level of the input signal stays fixed. Thus, at the output, the modulation index is

$$m = \frac{3.2\%}{5} = 0.64\%.$$

(b) From (a),

$$m = 0.032,$$

$$m = \frac{\Delta f}{f_m} = \frac{\Delta f}{10 \text{ MHz}},$$

$$\Delta f = 0.32 \text{ MHz}.$$

(c) Parts (a) and (b) concern AM and FM transfer from the LO to the desired signal in the IF. This part concerns mixing between LO components, something that occurs regardless of the presence of a desired signal. The LO (strong signal) is at 100 MHz. The signal to mix with it is at 110 MHz. The difference frequency at 10 MHz occurs in the IF. The conversion loss (10 dB) tells how much of the signal amplitude (-20 dBm) is lost in mixing and the balance (-20 dB) tells how much further reductions can be expected because the signal entered by the LO port rather than the signal port. Thus the output level is

-20 dBm	signal
-10 dB	conversion
-20 dB	balance
-50 dBm	at 10 MHz in the IF

4.4 EFFECT OF MODULATION OF A MULTIPLIED SIGNAL

When the frequency of a frequency-modulated signal is multiplied, its frequency deviation Δf is multiplied by the same factor, but the modulation frequency f_m does not change. This is illustrated in Fig. 4.13a. Since the sidebands are proportional to $\Delta f/f_m$, they increase in magnitude by the multiplication factor.

Amplitude modulation is attenuated by the multiplier if it is operated, as it often is, in the saturation region, where an increase in input power causes less-than-proportional increase in output power. However, whether this is the case depends on the operating point on a curve such as Fig. 4.14. FM to AM and AM to PM conversion can also occur in the multipler. AM to PM conversion can be especially noticeable because of the accentuation of the PM by the multiplication process. Because a multiplier is basically a nonlinear device, mixing products can also be produced if there are sidebands on the multiplied signal.

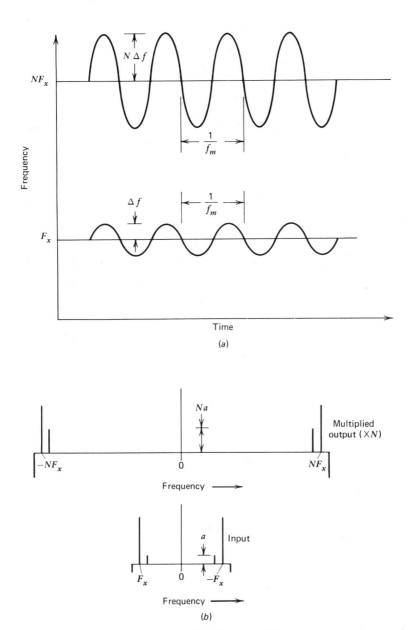

Figure 4.13 Frequency modulation on a multiplied signal: (*a*) instantaneous frequency and (*b*) Fourier spectrum.

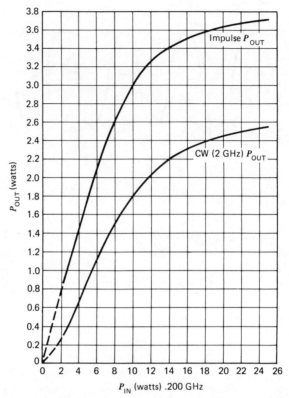

Figure 4.14 Power output versus power input for a step-recovery diode. Output is shown both without (impulse) and with (CW) output filter. From Hamilton, Fig. 19, p. 76. Reprinted from Microwave Journal by permission.

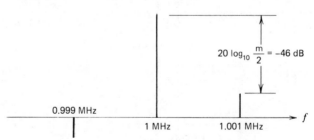

Figure 4.15 Input spectrum for Example 4.4.

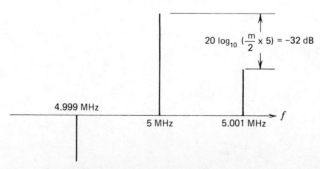

Figure 4.16 Spectrum of ×5 output in Example 4.4.

EXAMPLE 4.4

Problem A 1-MHz signal is frequency modulated at 1-kHz rate with a peak deviation of 10 Hz. What are the sideband frequencies and magnitudes after frequency multiplication by 5?

Solution At the multiplier input, we have

$$m = \frac{\Delta f_{peak}}{f_m} = \frac{10 \text{ Hz}}{1000 \text{ Hz}} = 10^{-2}.$$

See Fig. 4.15 for the input spectrum and Fig. 4.16 for the output spectrum.

4.5 EFFECT OF MODULATION OF A DIVIDED SIGNAL

A digital divider should not transmit amplitude modulation and, if it is biased for maximum sensitivity and driven hard enough, it should largely ignore AM. This is because it responds to zero crossings (crossings of the average value) of the input signal and these are unaffected by AM. However, if it switches at other than the zero crossings, AM on a driving sine wave will be converted to PM, at the same modulation frequency, at the output of the divider. This is shown in Fig. 4.17. AM on rectangular driving waveforms has even less effect. Phase (or frequency) modulation on the input to a digital divider causes phase modulation on the divider output. Consider a frequency divider driven by a slowly modulated wave. Each output transition is synchronized with an input

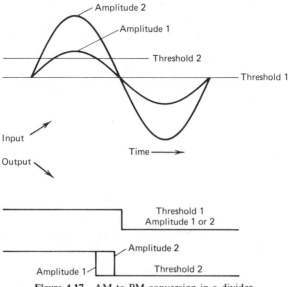

Figure 4.17 AM to PM conversion in a divider.

transition. Therefore, any time modulation of that input transition causes the same time modulation of the corresponding output transition. However, the time modulation at the output represents $1/N$ times the phase modulation that the same amount of time modulation represents at the input. This is because the period of the output is N times longer. That is, at the input, the phase change, in cycles, is

$$\Delta\varphi_{ic} = \frac{\Delta t}{T_i} = f_i \Delta t. \tag{4.9}$$

where Δt is the change in the time of the transition, T_i is the period, and f is the frequency. At the output, the phase change is

$$\Delta\varphi_{oc} = \frac{\Delta t}{T_o} = f_o \Delta t \tag{4.10}$$

$$= \frac{f_i}{N}\Delta t = \frac{\Delta\varphi_{ic}}{N}, \text{Q.E.D.} \tag{4.11}$$

The modulation sidebands will, therefore, be smaller by N at the output. Since extremes of deviation will occur simultaneously at the input and output, the modulating frequency will not change.

EXAMPLE 4.5

Problem What is the spectrum of the output of a frequency divider that divides the input signal of Example 4.4 by 5?

Solution From Example 4.4, m is 10^{-2} at the input. Figure 4.18 shows the output spectrum near the fundamental; of course there are harmonics also in the spectrum of a rectangular output voltage such as is generated by a digital divider.

While divider output transitions are synchronized with input transitions, there is always a finite delay T_D. This produces a phase shift ωT_D that must be included in the loop transfer function for stability analysis, or for determination of loop response, unless the delay is much smaller than the reciprocal of the loop bandwidth ω_L, (say $T_D < 1/(50\omega_L)$). Often this inequality is satisfied, allowing the delay through the divider to be ignored.

In order to further study the effects of phase modulation in dividers, let us

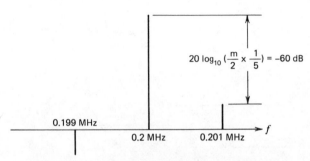

Figure 4.18   Spectrum of +5 output for Example 4.5.

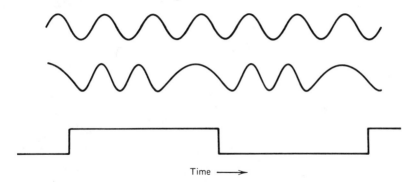

Figure 4.19   Two inputs which produce the same divider output.

consider how phase modulation might be more specifically defined at the divider input and output. At the input, we assume a constant-amplitude sinusoidal driving signal, whose phase can be easily determined from its instantaneous value by reference to an unmodulated wave. If the output phase is defined as $1/N$ times the input phase, all transitions of either polarity will be separated by one output cycle, as they should be, and the phase of the output from a divider which is driven by an unmodulated signal will be linear with time. Even though the phase of a divider output can be defined for each part of its waveform, the degree of phase modulation, which is measured by comparing modulated and unmodulated waveforms, can be observed only at a transition of the modulated output because the modulated output has an identifiable phase only at a transition. Only at transitions can the output indicate its phase and only at transitions can it affect a system. Input modulation occurring between every Nth zero crossing has no effect. This is illustrated in Fig. 4.19. Therefore, there is essentially a sampling process[3] occurring; phase information is only transmitted through the divider at the transition time.

Often the phase of the output of a divider is measured by a device (phase detector) which responds to only one transition, say the negative-going transition. The phase at this time is defined in accordance with the above discussion as illustrated in Fig. 4.20. Serious analytical difficulty can be caused, however, by the fact that the sampling that occurs at this transition is time-modulated and signal-dependent. Therefore, an approximation is made wherein the phase of the divider output (defined as $1/N$ times the input phase) is sampled at the negative transition of the reference (unmodulated) output. This is shown in Fig. 4.21 where $\Delta\varphi_0$ serves as an approximation for $\Delta\varphi_1$. The error is small if the peak frequency deviation is small compared to the frequency so that the slopes of the two ramps in Fig. 4.21 are approximately equal.

In the sampling representation, the sampled phase is represented by the area of an impulse occurring at the instant of sampling. If these pulses are low

3   Bracewell, pp. 188–194.

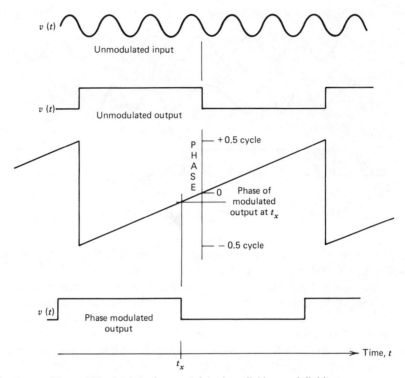

Figure 4.20 Input and output phase in a divider—a definition.

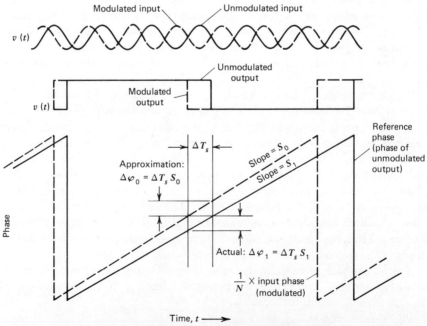

Figure 4.21 Constant-sampling-rate approximation.

passed, as will effectively occur for dividers in phase-locked loops, the impulse sampling will have no effect at low modulation frequencies. But when the phase-modulation frequency approaches the divider-output frequency, or a multiple thereof, the effect is significant. Since the sampling process causes the frequency spectrum to be repeated at frequency intervals equal to the sampling frequency, a modulation frequency which is separated from a multiple of the divider output frequency by Δf will cause modulation of the divider output at Δf. This is illustrated in Fig. 4.22. Thus, for example, if a phase-locked

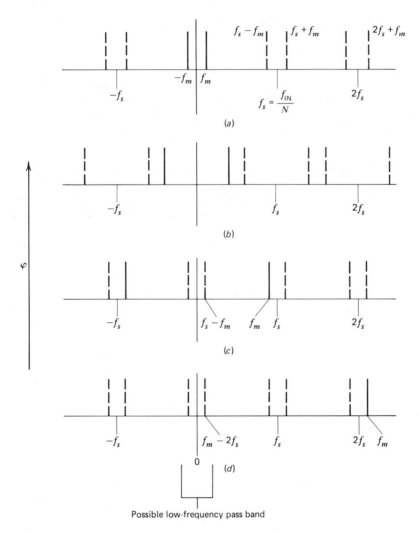

Possible low-frequency pass band

Solid lines: original and sampled spectrum   Dashed lines: sampled spectrum only

Figure 4.22 Creation of low-frequency modulation due to sampling, spectrum of phase deviation at the divider output. The original modulation increases in frequency from (a) to (d), but low-frequency components are created by sampling.

synthesizer has a divider with a 10-kHz output, and if a 50.1-kHz spurious signal is picked up at the divider input, the divider output will be phase modulated at a 100-Hz rate. While 50.1 kHz is much higher than the loop bandwidth and would be highly attenuated, 100 Hz may be well within the loop bandwidth and could cause objectionable FM on the synthesizer output. The resultant phase deviation at a 100-Hz rate is equal in strength to the original phase deviation at 50.1 kHz and the VCO will attempt to produce phase modulation at a 100-Hz rate to cancel the effect of the 50.1-kHz modulation at the divider output.

While the effect may differ somewhat for different types of phase detectors, it seems that sampling is inherent in the frequency division process, wherein phase information is transmitted only at output transitions. Frequency translations, similiar to those discussed above, are therefore to be expected.

The detailed spectrum of the divider output voltage may be obtained from pulse-modulation theory for pulse-position modulation with natural sampling.[4] However, this will not usually be needed for synthesizer work.

EXAMPLE 4.6

Problem   The output of a phase-locked synthesizer is mixed down to 100 MHz and then divided by 10,000. As a result of the mixing process, a 29-kHz signal, 30 dB weaker than the desired signal, also appears at the divider input. The divider is insensitive to AM at its input. The divider output drives a phase detector whose output is low passed at 1.5 kHz. What voltage appears at the output of the low-pass filter if it is lossless and the phase detector produces 1 V/rad ? What is the frequency of the voltage?

Solution   The decomposition of the single sideband to AM and FM is shown in Fig. 4.23. We ignore the AM sidebands because the divider is not sensitive to them. The remaining sidebands represent a peak phase deviation of 30 dB below 1 rad at a modulation rate of 99.971 MHz. The divider output is at 10 kHz and the peak phase

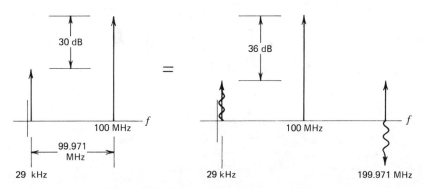

Figure 4.23   Decomposition for Example 4.6.

4   Westman, p. 21-18.

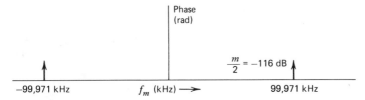

Figure 4.24 Spectrum of phase deviation at divider output without sampling, for Example 4.6.

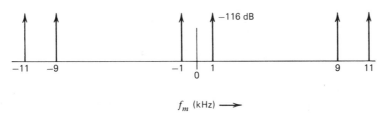

Figure 4.25 Spectrum of phase deviation at divider output including the effects of sampling, for Example 4.6.

deviation there is smaller by 10,000, thus 110 dB below 1 rad. The modulation frequency has not changed, however, so the spectrum of the phase deviation would be as shown in Fig. 4.24, except that the phase is sampled at a 10-kHz rate. Sampling at 10 kHz produces a spectrum consisting of copies of the original spectrum shifted by all possible multiples of ± 10 kHz. This produces the spectrum shown (at low frequencies) in Fig. 4.25. This phase is converted to a voltage in the phase detector and only components below 1.5 kHz are passed unattenuated through the filter. The only component thus passing results from 1 kHz at 110 dB below 1 rad peak. Note that a cosine with unity peak value has Fourier components of magnitude $\frac{1}{2}$ at positive and negative frequencies. Thus, -116-dB sidebands represent -110-dB peak waveform. Since the phase-detector transfer constant is 1 V/rad, its output is at 110 dB below 1 V, or 3.16 μV, peak, at 1 kHz.

4.6 OSCILLATOR SPECTRUMS

The frequency of a synthesizer is ultimately derived from the frequency of one or more oscillators by the processes which we have discussed. Therefore, the frequency stability of the synthesizer output depends on the stability of oscillators; the characteristics of the synthesizer's spectrum, its purity, are dependent upon those of the oscillators used. For this reason, it is important that we understand the general characteristics of the spectrums of oscillators and the parameters used to describe the spectrums of both oscillators and synthesizers.

An oscillator modifies the noise that is present in its electronic circuitry.

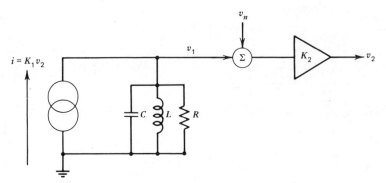

Figure 4.26   Simplified oscillator.

We will be able to show, by a relatively simply development,[5] the relationship of circuit noise to oscillator noise.

Figure 4.26 shows a simplified schematic of an oscillator. In normal operation, the nonlinearities in the circuit cause K_1 or K_2 to have a value such that the loop gain is unity, a condition for stable oscillation. The other condition is that there be 360° phase shift around the loop. This occurs at the resonant frequency of the tuned circuit where i_1, v_1, and v_2 are all in phase. What happens, then, when a small noise voltage v_n is introduced? In the absence of feedback ($K_1 = 0$), the noise will appear at the output, amplified by K_2. But what effect does the feedback have?

At frequencies well removed from the passband of the tuned circuit, the feedback becomes unimportant due to its small magnitude. Thus, far from spectral center the output contains amplified circuit noise. Close to center, however, the feedback has an important effect. In this region, based on an assumption that v_n is random, we can consider the noise power to consist one-half of AM and one-half of PM sidebands on the main signal.[6] While AM can be accentuated by the feedback, it also should be partially suppressed by the nonlinearity of the circuit. In any event, it is generally smaller than the PM near the spectral center[7] and may have less effect on systems due to limiting in system components. We therefore concentrate on the manner in which PM is accentuated by the oscillator loop.

In Fig. 4.26, one factor in the open loop gain is given by

$$\frac{v_1}{i} = Z_T = \frac{1}{1/R + j\left[\omega C - 1/(\omega L)\right]}. \tag{4.12}$$

The phase shift of this factor is

$$\Theta_{Tr} = \tan^{-1}\left\{ R\left[\omega C - 1/(\omega L)\right]\right\}. \tag{4.13}$$

5   Leeson, "'A Simple Model . . . ."
6   Taub.
7   Johnson.

At the resonant frequency ω_{osc} the rate of phase change with frequency may be obtained from Eq. (4.13) as

$$\left.\frac{d\Theta_{Tr}}{d\omega}\right|_{\omega_{\text{osc}}} = 2RC \tag{4.14}$$

$$= \frac{2Q}{\omega_{\text{osc}}}, \tag{4.15}$$

where Q is the loaded quality factor of the frequency-determining circuit. Thus for modulation rates small compared to one-half of the resonant bandwidth of Z_T, a change in frequency of $\delta\omega$ causes a phase change of $\delta\Theta_r$ in the transfer function, where the two changes are related by Eq. (4.15):

$$\delta\omega \approx \frac{\omega_{\text{osc}}}{2Q}\delta\Theta_r. \tag{4.16}$$

In order to maintain the required zero phase shift around the loop in the presence of modulation due to noise, the loop must respond to a phase perturbation of $\delta\Theta_{nr}$ by producing an equal and opposite phase change, and this is done by shifting the frequency by an amount given by Eq. (4.16). In other words, a phase modulation with peak deviation $\delta\Theta_{nr}$, due to noise v_n, necessitates frequency modulation of the oscillator output with peak deviation given by

$$\delta\omega_n \approx \frac{\omega_{\text{osc}}}{2Q}\delta\Theta_{nr}. \tag{4.17}$$

This, in turn, implies a resulting peak phase modulation of the oscillator output [see Eq. (4.4)] of

$$\delta\Theta_{2r} = \frac{\delta\omega_n}{\omega_m} \tag{4.18}$$

$$= \frac{\omega_{\text{osc}}}{2Q\omega_m}\delta\Theta_{nr}, \tag{4.19}$$

where ω_m is the noise modulation frequency. The process is pictured in Fig. 4.27. Thus the phase noise that would ordinarily appear at the output due to circuit noise is accentuated by a factor $f_{\text{osc}}/2Qf_m$. This is illustrated in Fig. 4.28, where white phase noise is shown with and without the influence of the oscillator loop. Here the noise has a continuous spectrum and the spectral density of phase is S_φ in units of rad^2/Hz bandwidth.

Additional filtering of the output may take place if the tuned circuit is between the output and the noise source.

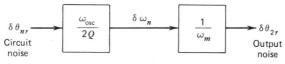

Figure 4.27 Conversion of noise in an oscillator.

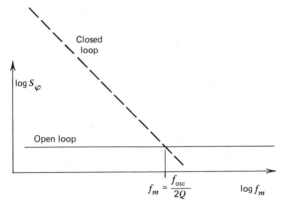

Figure 4.28   Phase noise in an oscillator.

4.7   RELATIONSHIP BETWEEN SPECTRAL DENSITIES[8]

We may consider the oscillator noise sidebands to consist of closely spaced discrete sidebands, even when, in reality, they are part of a continuous spectrum, by choosing the spacing close enough so the discrete sidebands would have the same effect in the problem being analyzed. If we choose a uniform spacing of δf, then the discrete equivalent to the continuous noise-power-density spectrum at frequency f_i, $p(f_i)$, has a sideband of power equal to

$$P_i = p(f_i)\delta f \tag{4.20}$$

located at frequency f_i. Here $p(f_i)$ is expressed in units of W/Hz and P_i would be in watts. P_i is the total power in components with frequency between $f_i - \frac{1}{2}\delta f$ and $f_i + \frac{1}{2}\delta f$ if $p(f_i)$ is the "one sided" power spectral density. (The Fourier representations employs negative and positive frequencies and each contains only one-half of the total power. This is called a "two-sided" representation and is the Fourier transform of the autocorrelation function. We will use one-sided densities unless specified otherwise.) For example, if the power density at 100 Hz is 1 mW/Hz, the power in a 0.2-Hz band there will be 0.2 mW.

The term power spectral density is also applied to voltage, that is,

$$\tilde{V}_i^2 = \tilde{v}^2(f_i)\delta f, \tag{4.21}$$

where $\tilde{v}^2(f_i)$ is the mean square (squared rms) voltage density in V²/Hz. Similarly, the term may be applied to the density of phase (noise), that is

$$\tilde{\varphi}_i^2 = S_\varphi(f_i)\delta f, \tag{4.22}$$

8 Howe.

where $S_\varphi(f_i)$ is the power spectral density of phase, at modulation frequency f_i, and $\tilde{\varphi}_i^2$ is the square of the rms phase deviation for frequencies from $f_i - \frac{1}{2}\delta f$ to $f_i + \frac{1}{2}\delta f$. The units for $\tilde{\varphi}_i^2$ and S_φ are rad^2 and rad^2/Hz or cycles^2 and $\text{cycles}^2/\text{Hz}$, respectively. For example, in some region where S_φ is 4 cycles^2/Hz, the square of the rms phase deviation in a 2-Hz band is 8 cycles^2. Sometimes the square root of S_φ may be given, that is, in this example, 2 cycles$/\sqrt{Hz}$.

Passage of a signal through a phase detector produces a voltage power spectrum from a phase power spectrum and the two are related by the phase detector's gain constant K_φ. Thus if a sinusoidal phase modulation with peak deviation of m_i caused a peak sinusoidal voltage of V_i from a phase detector, the transfer function would be

$$K_\varphi = \frac{V_i}{m_i}. \tag{4.23}$$

Squaring this expression, we can obtain the mean square output,

$$\tilde{V}_i^2 = K_\varphi^2 \tilde{\varphi}_i^2. \tag{4.24}$$

If $\tilde{\varphi}_i$ is the rms phase deviation in a differential bandwidth at frequency f_i, the phase (power) spectral density $S_\varphi(f_i)$ would then cause a mean square output voltage density given by

$$\tilde{v}^2(f_i) = K_\varphi^2 S_\varphi(f_i). \tag{4.25}$$

EXAMPLE 4.7

Problem A peak phase deviation of 1 rad causes a peak voltage of 0.1 V from a phase detector. How much noise density at 1 kHz will appear at the phase-detector output due to an input which has $S_\varphi(1 \text{ kHz}) = 0.2 \text{ rad}^2/\text{Hz}$?

Solution
By Eq. (4.23),

$$K_\varphi = \frac{0.1 \text{ V}}{1 \text{ rad}} = 0.1 \text{ V/rad}.$$

By Eq. (4.25),

$$\tilde{v}^2(1 \text{ kHz}) = (0.1 \text{ V/rad})^2 (0.2 \text{ rad}^2/\text{Hz})$$

$$= 0.002 \text{ V}^2/\text{Hz}.$$

Similarly, rms frequency deviation $\tilde{\Delta}\omega$ can be represented by a density, where

$$\tilde{\Delta}^2\omega_i = S_{\varphi r}(f_i)\delta f_i. \tag{4.26}$$

A similar density can be defined for AM. Some of these relationships are illustrated in Fig. 4.29. Note that only *power* densities are added. Only densities with units of V^2/Hz, rad^2/Hz, Hz^2/Hz, etc. (not $\text{V}/\sqrt{Hz}$, $\text{rad}/\sqrt{Hz}$, nor $\text{Hz}/\sqrt{Hz}$) can be added.

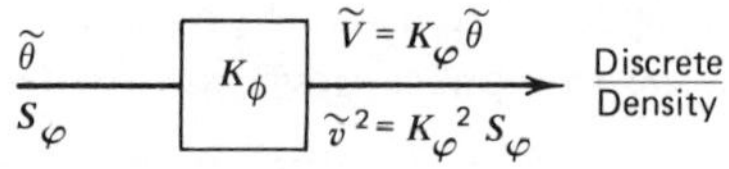

Figure 4.29 Densities and discrete signals.

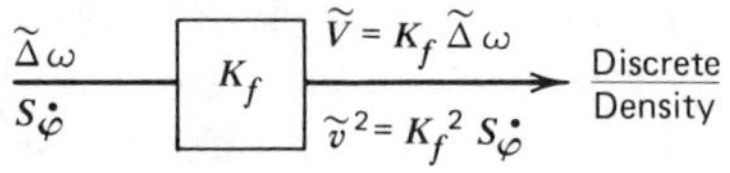

Using Eqs. (4.21), (4.22), and (4.26), which relate densities to discrete signals, the relationships between phase and frequency deviations and sideband levels given in Section 4.1 for small modulation indices can be extended to densities. Thus, an FM-deviation density of $S_{\dot{\varphi}r}(f_m)$ in units of $(\text{rad}/\text{sec})^2/\text{Hz}$ implies a phase spectral density of

$$S_{\varphi r}(f_m) = \frac{S_{\dot{\varphi}r}(f_m)}{\omega_m^2} \tag{4.27}$$

and a relative single-sideband level for the voltage spectrum separated by f_m from spectral center is

$$\mathcal{L}(f_m) \triangleq \frac{S_{v1}(f_m)}{\tilde{V}_o^2} = \frac{S_{\varphi r}(f_m)}{2}, \tag{4.28}$$

where S_{v1} is sideband density of mean square voltage and $\tilde{V}_o^2$ is the total mean square voltage. This says that, for small-modulation-index FM, the sideband level of the voltage spectrum has the same shape as the phase spectral density, and that the relative power density, at a separation from spectral center of f_m, is one-half of the phase spectral density, expressed in rad^2/Hz, at a separation of f_m. What is more, the total relative spectral power density due to FM at f_m from center (double sideband) equals $S_\varphi(f_m)$. These relations are illustrated in Fig. 4.30.

The statement is generally made that this relationship between phase spectrum and voltage spectrum is valid for total rms phase deviation much smaller than 1 rad.[9] The latter would be obtained by integrating the area under the phase spectral density curve and taking the square root of the result. The objective is to allow the small-modulation-index approximations to remain valid. This restriction may sound reasonable at first, but when one considers that typical oscillators have phase-noise spectra that climb rapidly (-9 dB/octave slope) at lower and lower frequencies—in crystal oacillators, this apparently extends to modulation rates with periods exceeding a year[10]—

9 Howe, p. 5.
10 Allen.

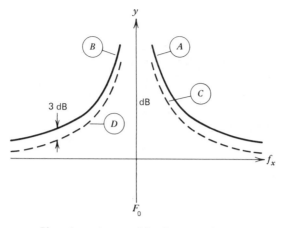

Phase (power) spectral density, curve A:

$$y = 10 \, log_{10} \, S_\varphi \text{ (in rad}^2/\text{Hz)}$$

$$F_0 = 0, \quad f_x = f_m;$$

DSB spectral density, curve A:

$$y = 10 \, log_{10} \, S_{v2}/\tilde{V}_o^2,$$

$$F_0 = 0, \quad f_x = |F - F_C|;$$

SSB spectral density, curves C and D:

$$y = 10 \, log_{10} \, S_{v1}/\tilde{V}_o^2 = 10 \, log_{10} \, \pounds$$

$$F_0 = F_C, \quad f_x = F.$$

Figure 4.30 Comparison of density plots.

this restriction seems more severe. What keeps this restriction from destroying the practical value of the relationship is that:

1. We are often not interested in the lowest modulation frequencies. The importance of modulation with a period of a day or a week is surely diminished if the signal is only observed for a second.

2. If the integrated phase spectral density at frequencies greater than f' is small, then only the modulation frequencies well below f' will cause a broadening of the "carrier" and frequencies beyond f' can be considered to modulate this broadened carrier. This will lead to some "smearing" of the spectrum, but it will not be noticeable at offsets beyond f', where the spectral power density does not change much over any band with width equivalent to the broadened carrier. Therefore, a criterion for acceptance of the small-modulation-index approximation [Eq. (4.28)] for modulation frequencies higher than f' is[11]

$$\int_{f'}^{\infty} S_\varphi(f_m) \, df_m \ll 1 \text{ rad}^2. \tag{4.29}$$

11 Shoaf, p. 43.

4.8 TYPICAL SPECTRAL SHAPES

In the absence of feedback, an oscillator circuit will produce noise as shown in Fig. 4.31. Thermal noise density (kT) is multiplied by the active-device transducer power gain (G) and noise figure (F) to produce a white-noise level.* At low frequencies the noise climbs with a -3-dB/octave slope. This is called $1/f$ noise because its power spectral density is proportional to $1/f$. It is also called flicker noise.

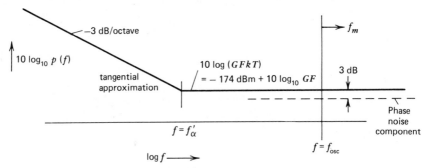

Figure 4.31   Oscillator circuit noise with loop open.

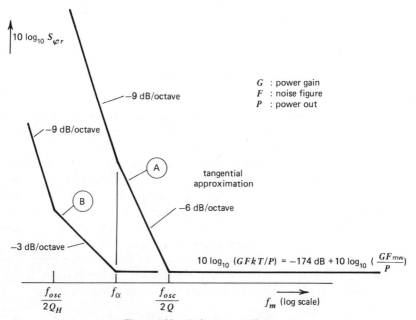

Figure 4.32 S_ϕ for two oscillators.

*T is $290°K$ if F includes the effect of temperature. Transducer gain = (power to actual load)/(availabe power).

When the effect of oscillator feedback is included, the additive phase noise increases in magnitude at frequencies closer than $f_{osc}/(2Q)$ to the spectral center, as described above, taking on a -6-dB/octave slope as shown in curve A of Fig. 4.32. Apparently because at least some types of $1/f$ noise modulate the current flow in a transistor,[12] noise of this type, originating at a frequency f_m, can cause modulation of the signal at a modulation frequency of f_m. (A similar mechanism apparently exists in other amplifying devices.) Therefore, below some small frequency offset from center, an additional -3-dB/octave slope occurs. For very-high-Q oscillators, for example, crystal oscillators, it is common for the increase in noise due to $1/f$ modulation to occur at a higher frequency separation than $f_{osc}/(2Q)$, and, therefore, the shape shown in Fig. 4.32, curve B, occurs. Note that the level of the noise floor in Fig. 4.32 is the same as the ratio of the noise power density to total oscillator power in Fig. 4.31. This is because of two counteracting factors: while the noise floor in Fig. 4.32 represents only the phase noise, which is one-half of the noise floor in Fig. 4.31, S_φ equals the sum of both sidebands from Fig. 4.31.

EXAMPLE 4.8

Problem Sketch the expected room-temperature spectrum for an oscillator with the following parameters:

f_{osc}, 10 GHz;
Q, 200;
f_α, 10 kHz;
F, 7 dB;
G, 10 dB;
P, 10 dBm;

Solution The noise floor is obtained from

$$S_\varphi\left(f > \frac{f_{osc}}{2Q}\right) = -174\ \text{dB} + 10\ \log_{10}\left(\frac{GF\ \text{mW}}{P}\right)$$

$$= -174\ \text{dB} + 10\log_{10}G + 10\log_{10}F - 10\log_{10}\left(\frac{P}{\text{mW}}\right)$$

$$= -174\ \text{dB} + 10\ \text{dB} + 7\ \text{dB} - 10\ \text{dB} = -167\ \text{dB}.$$

The spectrum is sketched in Fig. 4.33. Note the following relationships used in producing the sketch:

$$3\ \text{dB/octave} = 10\ \text{dB/decade} \rightarrow 10\log_{10}f,$$
$$6\ \text{dB/octave} = 20\ \text{dB/decade} \rightarrow 20\ \log_{10}f,$$
$$9\ \text{dB/octave} = 30\ \text{dB/decade} \rightarrow 30\log_{10}f.$$

By examination of Fig. 4.32, we can see which oscillator parameters influence spectral purity. Far from spectral center phase noise can be reduced by maintaining a high power level at the input to the active device (P/G) and

12 Pritchard.

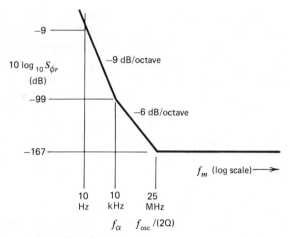

Figure 4.33   Sketch of spectrum for Example 4.8.

using a device with a low noise figure (F). Low phase noise far from center can be important in an LO because, when such noise is transferred to a strong incoming signal, it can overpower a weak signal in the receiver IF. This is illustrated in Fig. 4.34. The process is called (one form of) receiver desensitization. The process is essentially the same when the noise sidebands are on a transmitted signal. Desensitization due to an LO can be improved in a receiver through front-end filtering, but this will not help if the sidebands are on an interfering transmitted signal.

Phase noise can be reduced closer to spectral center by the same means as noted above, and also by increased Q and by the use of an active device with lower $1/f$ noise.

Other types of noise may also be observed, particularly increased levels very near center [13] and microphonic responses to vibration.

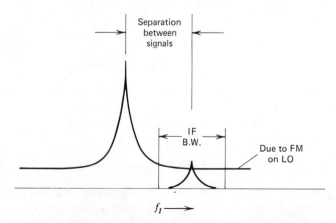

Figure 4.34   Desensitization of a receiver by LO noise.

13   Howe, p. 13.

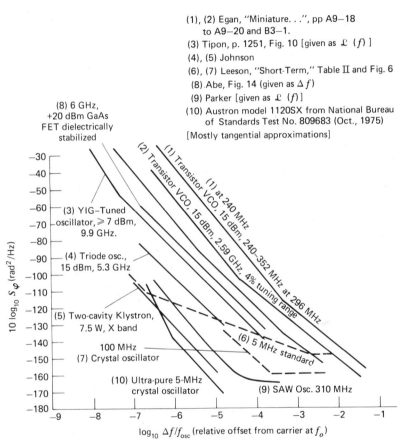

(1), (2) Egan, "Miniature. . .", pp A9–18
    to A9–20 and B3–1.
(3) Tipon, p. 1251, Fig. 10 [given as $\mathcal{L}$ (f)]
(4), (5) Johnson
(6), (7) Leeson, "Short-Term," Table II and Fig. 6
(8) Abe, Fig. 14 (given as Δf)
(9) Parker [given as $\mathcal{L}$ (f)]
(10) Austron model 1120SX from National Bureau
    of Standards Test No. 809683 (Oct., 1975)
[Mostly tangential approximations]

(8) 6 GHz,
+20 dBm GaAs
FET dielectrically
stabilized

(2) Transistor VCO, 15 dBm, 2.59 GHz, 4% tuning range

(1) Transistor VCO, 15 dBm, 240–352 MHz at 296 MHz

(1) at 240 MHz

(3) YIG–Tuned
oscillator, ≥ 7 dBm,
9.9 GHz.

(4) Triode osc.,
15 dBm, 5.3 GHz

(5) Two-cavity Klystron,
7.5 W, X band

100 MHz
(7) Crystal oscillator

(6) 5 MHz standard

(10) Ultra-pure 5-MHz
crystal oscillator

(9) SAW Osc. 310 MHz

$10 \log_{10} S_\varphi$ (rad^2/Hz)

-30
-40
-50
-60
-70
-80
-90
-100
-110
-120
-130
-140
-150
-160
-170
-180

-9 -8 -7 -6 -5 -4 -3 -2 -1

$\log_{10} \Delta f/f_{osc}$ (relative offset from carrier at f_o)

Figure 4.35 Some oscillator FM noise curves.

Note that, because of the appearance of f_{osc} in the expression for the noise break point, close-in noise sideband levels tend to be proportional to oscillator frequency. One other factor of importance is oscillator tuning range. While not appearing explicitly in the equations determining noise level, wide tuning range often lessens the loaded Q of varactor-tuned oscillators, causing increased noise levels. YIG oscillators, however, have inherently wide tuning range and can be made with high Q.[14]

Phase-noise spectra for many oscillators have been reported in the literature.[15, 16] Figure 4.35 shows some such curves. Note how normalization of the abscissa in the manner shown tends to make such curves relatively independent of oscillator frequency (f_{osc}). Note also how high power and high Q (YIG, SAW devices, cavitites, and crystals) reduce noise and how wide bandwidth can increase it.

4.9   COMPONENT NOISE

The analog circuits between the divider output and the VCO input are all noise sources. To assure a desired level of spectral purity, the designer must become familiar with the noise sources contained in active and passive devices[17] and analyze the effect of each on the synthesizer spectrum. While a study of these noise sources is beyond the scope of this work, we note, in general, that both resistors and semiconductor devices possess flicker $(1/f)$ noise which has a level higher than white noise at the lowest frequencies. Flicker noise on the tuning line causes -9-dB/octave spectral density on the VCO output while white noise causes -6-dB/octave density. Since these slopes are also produced by the oscillator itself, they can be difficult to distinguish in an operating synthesizer.

4.10   NOISE DENSITY AT THE SYNTHESIZER OUTPUT

The synthesizer output variable f_{OUT} equals the loop phase error multiplied by K_F, the forward gain at ω equal to one, plus added frequency changes caused by noise. These latter may enter the control system at f_1 in Fig. 3.7a. We know how the output frequency reacts to frequency changes at f_1; this is shown in Fig. 3.7b for a simple loop. The transfer function for phase is the same as for frequency, since

$$\frac{f_{OUT}}{f_{IN}} = \frac{s\varphi_{OUT}}{s\varphi_{IN}} = \frac{\varphi_{OUT}}{\varphi_{IN}}. \tag{4.30}$$

14   Tipon.
15   Leeson, "Short-term ... "
16   Johnson.
17   Motchenbacher.

Noise Density at the Synthesizer Output

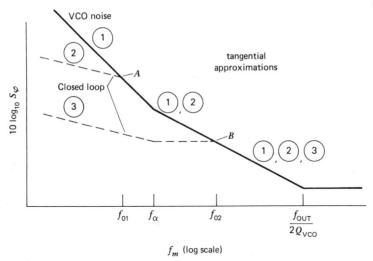

Figure 4.36 Output noise density due to VCO.

With the loop open, the VCO phase noise, which enters at f_1, appears at f_{OUT}, as illustrated in Fig. 4.36, curve 1. With the loop closed, the noise entering at f_1 is multiplied by the closed-loop transfer function $f_{OUT}(\omega)/f_1(\omega)$, as shown in Fig. 3.11 for a highly damped loop. This results in a modification of the output spectrum, as illustrated for two different values of $f_0 \, (= \omega_0/2\pi)$, f_{01}, and f_{02}, by curves 2 and 3 in Fig. 4.36. Thus, the spectrum is improved at offsets less than f_0. For lower damping factors, there would be peaking at the corner where the VCO noise flattens (points A and B in Fig. 4.36) and these peaks would move to lower frequencies.

Frequency changes due to <u>reference noise</u> enter at f_{REF} in Fig. 3.7a and are transferred as shown by Fig. 3.7c and Fig. 3.11. Figure 4.37 illustrates a typical reference-oscillator phase-noise profile $S_{\varphi,\,REF}$ and the result of multiplication by the transfer function f_{OUT}/f_{REF}, as shown in curves 1 and 2.

The effect of noise entering at another place in the loop can be analyzed by writing a closed-loop response from that point to the output, but it is usually simpler to relate such noise to noise at f_{REF} or at f_1. For example, noise voltage appearing on the phase-detector output can be related to the equivalent reference phase noise by the phase-detector transfer constant K_φ. The loop output noise can then be obtained by multiplying this equivalent by the closed-loop transfer function from f_{REF} to f_{OUT}. Noise power density at the loop-filter output, p_{LF}, produces a VCO frequency deviation density given by

$$S_{\dot{\varphi}}(f_m) = K_v^2 p_{LF}(f_m) \tag{4.31}$$

and thus a phase deviation density,

$$S_{\varphi}(f_m) = \frac{K_v^2}{f_m^2}\, p_{LF}(f_m). \tag{4.32}$$

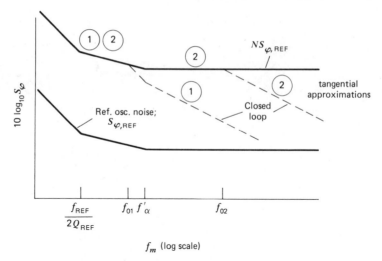

Figure 4.37 Output noise density due to reference oscillator.

These values can then be multiplied by f_{OUT}/f_1 to obtain the closed-loop output values.

Proper design of a phase-locked synthesizer requires consideration of the effect of all noise sources in the loop, both those generated in the loop components and those entering from outside sources, upon the output spectrum.

Figure 4.38a shows the open-loop phase-noise sidebands due to the VCO and to the multiplied reference oscillator. These cross at frequency f_x. A choice of loop parameters such that the responses f_{OUT}/f_1 and $f_{\text{OUT}}/Nf_{\text{REF}}$ are equal at this frequency produces a near-optimum design in the sense that $NS_{\varphi,\text{REF}}$ is attenuated above f_x, where it is greater, and $S_{\varphi,\text{VCO}}$ is attenuated below f_x, where it is greater. The resulting total noise is illustrated in Fig. 4.38b.

EXAMPLE 4.9

Problem   The reference for a loop with a divide number of 100 has a noise floor of -130 dB/Hz in the region of interest. The VCO has a -6 dB/octave noise slope in the same region and a noise density on this slope of -70 dB/Hz at 4 kHz from spectral center. What would be the optimum loop bandwidth for a simple loop, with no filter, to improve the output spectrum? What would be the total single-sideband power spectral density at 4 kHz after the loop is closed (use small-modulation-index approximation)?

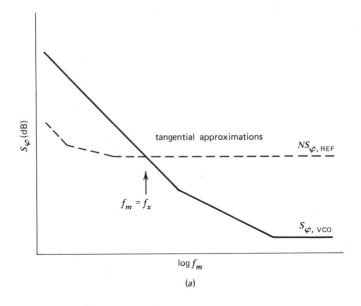

(a)

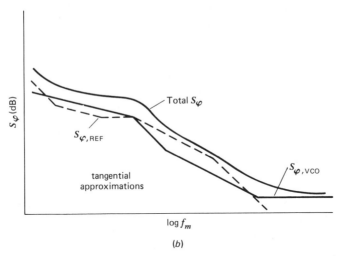

(b)

Figure 4.38 Effect of loop on synthesizer broadband output noise: (a) open loop and (b) closed loop.

Solution   The VCO noise and the multiplied reference noise densities are shown in Fig. 4.39. The crossover point f_x is at 40 kHz because, at one decade from 4 kHz, the VCO's phase noise has decreased 20 dB from -70 dB to equal the multiplied-reference phase-noise level. 40 kHz is the near-optimum loop bandwidth.

With the loop closed, at 4 kHz the multipled reference noise and the attenuated VCO noise are both at -90 dB/Hz, giving -87 dB/Hz for the sum.

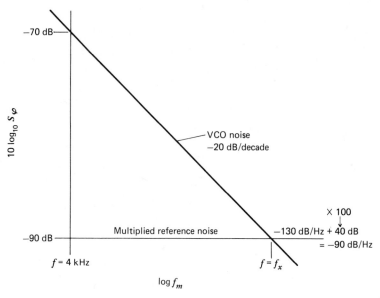

Figure 4.39 Phase noise density for Example 4.9.

PROBLEMS

4.1 A 10-dBm signal at 1 kHz and a −45-dBm signal at 1.2 kHz enter a limiting device which removes all amplitude modulation. The output contains a −10-dBm signal at 1 kHz. What other component(s) do you expect to find and at what amplitude(s)? List only the most significant.

4.2 A 30-MHz signal is frequency modulated at 10-kHz rate with a peak deviation of 10 Hz. (1) what are the first FM-sideband frequencies and magnitudes? (2) What are the sideband frequencies and magnitudes after frequency multiplication by 10? (3) After frequency division by 100?

4.3 A 1-MHz LO enters a mixer at +20 dBM along with a 1.02-MHz single spurious sideband at −20 dBm. The mixer conversion-loss curve (dB out versus dB in) has a slope of $\frac{1}{8}$ at the operating point. (1) How much AM (in percent modulation) appears on the mixer IF output? (2) How much peak frequency deviation? (3) If the mixer has 7-dB conversion loss and −25-dB unbalance for LO-port signals, what signal (magnitude and frequency) would be expected below 0.5 MHz at the mixer output?

4.4 A 20-MHz signal is divided to 10 kHz. A 7.1-kHz signal, 35 dB weaker than the desired signal, also appears at the divider input. The divider is insensitive to AM at its input. The divider output drives a sample-and-hold phase detector whose output is low passed by a filter that cuts off rapidly at 3.5 kHz. What is the frequency and voltage of the signal appearing at the output of the low-pass filter if it is lossless and the phase detector produces 1 V/cycle?

4.5 Show that the same IF level is obtained if (1) a single sideband on the LO is decomposed into AM and FM components and these mix with the LO as if (2) the SSB signal mixes directly with the LO.

4.6 Sketch the expected spectrums of oscillators with the parameters given below on log axes (i.e., dB vs. log f_m). Give the phase spectral density $S_{\varphi r}$ in rad^2/Hz, at 10-Hz modulation frequency (see Fig. 4.32).

	(a)	(b)	(c)
f_{osc}	1 kHz	100 MHz	10 MHz
Q	5	20	5000
f_α	250 Hz	2 kHz	4 kHz
F	20 dB	6 dB	6 dB
G	15 dB	10 dB	15 dB
P.	1 mW	10 mW	0 dBm

4.7 At $f_m = 30$ kHz, an oscillator has $S_{\varphi c} = 3$ Hz^2/Hz. What is the expected relative single-sideband power spectral density at 30 kHz from center in dB?

4.8 The reference for a loop with a divide number of 1000 has a phase-noise floor of -130 dB/Hz in the region of interest. The VCO has a -9-dB/octave phase-noise slope in the same region and a noise density on this slope of -32 dB/Hz at 1 kHz from spectral center. What would be the optimum loop bandwidth for a simple loop, with no filter, to improve the output spectrum? What would be the total single-sideband power spectral density at 4 kHz after the loop is closed (use small-modulation-index approximation)?

5

Phase Detectors

One of the most critical components in the indirect synthesizer is the phase detector. There are many types of phase detectors, each with its special characteristics. Some have the virtue of simplicity, or of allowing the rest of the design to be simpler. Some are useful at very high frequencies. Others have a very low content of undesirable signals in their output; this can easily be the difference between a design which is feasible and one which is not. We will discuss the properties of six different types of phase detectors which can be employed in frequency synthesis. We will also consider, in more detail than heretofore, how the phase-detection process affects the loop characteristics, for the effects of a not-too-obvious complication resulting from this process can be very severe.

5.1 TYPES OF PHASE DETECTORS

5.1.1 Balanced Mixer[1, 2]

We have already given some consideration to this type of phase detector and have noted that it is not appropriate for many synthesizer applications because of the high level of undesirable components. However, this is not always a severe disadvantage and the balanced mixer does receive considerably use.

The difference-frequency output of the mixer can be described as

$$v_d = A_d \cos(\varphi_1 - \varphi_2), \tag{5.1}$$

where

$$\varphi_i = \omega_i t + \theta_i. \tag{5.2}$$

Distortion of this characteristic is greatest when the two input levels are nearly

1 Kurtz.
2 Blanchard, pp. 8–29.

98

equal.[3, 4] In frequency-mixing applications, we are primarily interested in $\omega_1 - \omega_2$, but here we look upon φ_i as a phase, whether we consider it constant or not. The difference is really in how we consider the functioning of the balanced mixer; the operation is basically the same whether it is considered as a frequency mixer or a phase detector. Thus, while we concentrate on the difference-frequency or phase-difference output, given by Eq. (5.1), other signals also exist at the output. The strongest of these is the sum-frequency output, which is

$$v_s = A_s \cos\left[(\omega_1 + \omega_2)t + \theta_1 + \theta_2 \right]. \tag{5.3}$$

At phase lock, this has a frequency equal to twice the reference and, when A_s and A_d are equal, as they often are, a peak-to-peak amplitude equal to the total range of the desired phase-detector output, $2A_d$. Higher harmonics of the reference will also be present, but will be weaker.

A second significant output is at the reference frequency. Its magnitude is dependent upon the LO-to-IF isolation and the RF-to-IF isolation of the mixer. These can be as low as 20 dB, even in a doubly balanced mixer. While this signal is significantly weaker than the second harmonic of the reference, its lower frequency causes it to be a potentially significant problem.

Another signal which can cause significant problems is a dc component which is the result of rectification and imperfect balance in the mixer. This gives the output characteristic a dc offset. The problem is accentuated if the magnitudes of the phase-detector inputs are significantly different, because the desired phase-difference signal is weaker than the weakest of the two inputs, but the offset tends to be proportional to the total input power. This is not necessarily a strict proportionality, however, as the balance of the mixer can change as operating levels change. In some types of loops, acquisition of lock depends upon a very small error voltage, which is generated during cycle skipping, and the unbalanced rectification voltage could overcome this.

In spite of these undesirable features, the balanced mixer can be the best choice for a phase detector in some application, especially where the reference frequency is too high for some other phase detectors and where the loop bandwidth is sufficiently narrow, compared to the reference frequency, so that filtering can be effective. It also has less potential for problems due to the sampling of spurious signals, as discussed in the next section, than do the other types.

5.1.2 High-Speed Sampler

The high-speed sampler is a device which is very commonly used as a phase detector in locking a VCO to an integer multiple of a reference.[5] The concept of the sampler is illustrated in Fig. 5.1. This is basically a single-balanced

---

3 Krishnan.
4 Klapper, pp. 44–46.
5 Grove.

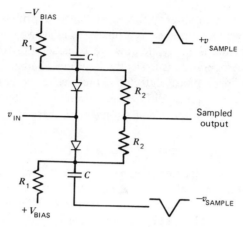

Figure 5.1   High-speed sampler.

mixer driven by a narrow pulse on the LO input. The pulse input, $2v_{SAMPLE}$, is balanced. During the pulse the two diodes conduct, causing the capacitors to charge to the input voltage less the pulse voltage. When the pulse returns to its quiescent value the average of the voltages at the two diode-capacitor interfaces remains equal to v_{IN} and this is the sampled output. A very narrow pulse, such as might be generated by a step-recovery diode, can be used to give good sampling of signals with frequencies as high as 12 GHz or more.

Figure 5.2a illustrates, in the frequency domain, how the phase detection occurs. The sampling process causes the input RF spectrum to be repeated at intervals equal to the sampling frequency.[6] As the frequency of the sampled

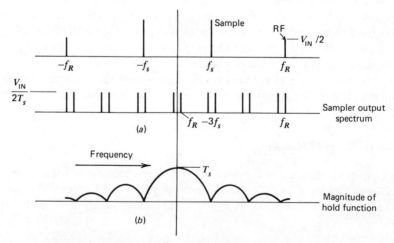

Figure 5.2   Spectral representation of sample-and-hold process.

6 Bracewell, pp. 188–194.

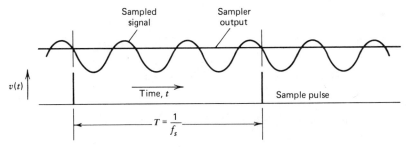

Figure 5.3   Phase detection by sampling.

signal approaches a multiple of the sample frequency, a low-frequency output appears and then moves lower in frequency, finally reaching zero frequency when the frequency of the sampled signal is an exact multiple of the sampling frequency. At that point, the sampler output voltage is proportional to the cosine of the phase difference between the sampled frequency and the corresponding harmonic of the sampling signal. This is shown in the time domain in Fig. 5.3.

The value of v_{IN} at each sample instant is held on the capacitors until the next sample instant. The impulse response of this process is represented as an integration of the impulse, whose area equals one, to produce a voltage step equal to one, followed by an equal but opposite step at the time of the next impulse. Thus, each impulse produces a voltage equal to one and lasting until the next impulse. The Laplace transform of this "zero-order hold" function is

$$G_s(s) = \frac{1}{s}(1 - e^{-T_s s}),\tag{5.4}$$

where T_s is the sample period. In terms of radian frequency, this is

$$G_s(\omega) = \frac{1}{j\omega}\left[1 - \exp(-j\omega T_s)\right]\tag{5.5}$$

$$= 2\frac{\exp(-j\omega T_s/2)}{\omega}\frac{\exp(j\omega T_s/2) - \exp(-j\omega T_s/2)}{2j}\tag{5.6}$$

$$= T_s\exp(-j\omega T_s/2)\frac{\sin(\omega T_s/2)}{\omega T_s/2}.\tag{5.7}$$

The Fourier transform of an impulse train in the time domain is an impulse train in the frequency domain. The amplitude of both trains is the same, except for a factor $1/T_s$ in the frequency domain. This is cancelled by T_s in the hold response [Eq. (5.7)]. The remaining function has a $(\sin x)/x$ shape, as shown in Fig. 5.2b, with nulls at multiples of the sampling frequency. Thus, at phase lock, only the dc component can be seen, as shown in the time-domain representation of Fig. 5.3. (That is not to say that there will not be leakage of the sampling frequency and its harmonics in a real sampler.) The phase shift

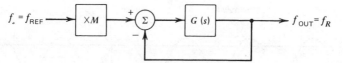

Figure 5.4   Block diagram for loop with high-speed sampler.

of the hold function is linear and is obtained from Eq. (5.7) as

$$\varphi_{sr} = \frac{2\pi f T_s}{2} \tag{5.8}$$

$$= \frac{\pi f}{f_s}. \tag{5.9}$$

Thus, at the sampling frequency f_s there is a 180° phase shift.

The low-frequency output from a sampler of this type can thus be used to phase lock an oscillator to a harmonic of a reference frequency. The block diagram for a loop where the oscillator is locked to the Mth harmonic of the reference frequency is shown in Fig. 5.4. The harmonic must be chosen by constraining the VCO to a range of frequencies which will insure lock at the desired harmonic.

One potential disadvantage of any sampling phase detector is that it can translate undesired signals to a lower frequency, where they are often more harmful. As can be seen from Fig. 5.2, the same process which translates f_R to zero frequency will translate any other frequency that is near some multiple of f_s to a low value. This can then pass through the hold function and other low-pass filtering and result in FM at the synthesizer output. The same thing occurs as a result of the effective sampling in a frequency divider, as was discussed in Section 4.5. As a result of this process, it is important that spurious signals not be introduced to the sampler along with the desired VCO output unless their frequencies are such that, after down conversion by a multiple of f_s, they can be sufficiently attenuated. The sampling process will also fold a broad noise floor over many times, translating the noise in the vicinity of many multipies of f_s to lower frequencies where it can pass through the loop.

An alternative way to obtain a system, such as is represented by Fig. 5.4, is by use of a times-M multipler and a balanced-mixer phase detector. While this may be more complex in some cases, it does avoid the translation to low frequencies of signals near multiples of f_{REF}, except for those near Mf_{REF} and multiples of Mf_{REF}. The latter are small compared to the desired signals, because the spurious responses of a balanced mixer are weaker than its difference (or sum) frequency.

The designer must keep this potential difficulty in mind whenever there is sampling. However, this is not to suggest that those very useful devices which employ sampling should be avoided, but only that caution should be exercised.

5.1.3 Exclusive OR[7]

An exclusive-OR gate can be used as a phase detector, as illustrated in Fig. 5.5. Figure 5.5*a* shows two square-wave inputs to the exclusive-OR gate of Fig. 5.5*b* at *A* and *B* and the resulting output waveform at *C*. The average value of the output is plotted in Fig. 5.5*c* as a function of the phase difference between inputs. This average value, the result of low passing the gate output, is proportional to phase over a range of 0.5 cycle. The ideal output shown does not contain any energy at the fundamental input frequency, regardless of phase difference. At a 90° phase difference, it has maximum second-harmonic content with peak-to-peak amplitude 1.27 times the peak-to-peak range of the phase-detector characteristic. If this output seems similar to that of the balanced mixer, it is because the balanced mixer operates as an exclusive-OR gate when driven by two rectangular waveforms of proper amplitude. Note

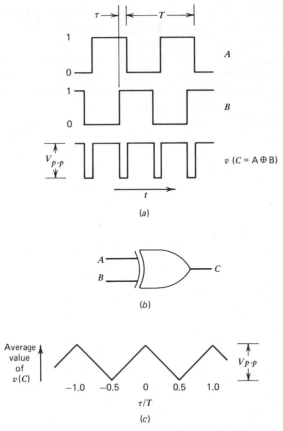

Figure 5.5 Exclusive-OR gate used as a phase detector.

7 Blanchard, pp. 24–28.

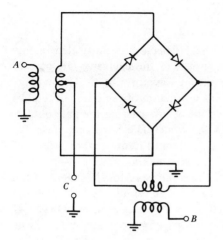

Figure 5.6   Balanced mixer.

that a voltage at A in Fig. 5.6 causes current to flow through either the left or right leg of the doide bridge, depending on the polarity of A. As the signal at A changes polarity, the conducting leg switches to the other side, and this inverts the polarity of a smaller signal at B as it appears at the output, C.

If either, or both, inputs has other than 50% duty cycle, the exculsive-OR phase-detector characteristic will have flat spots, as illustrated in Fig. 5.7.

The detailed harmonic contect of the output of this type, and of the following type, can be obtained as a function of duty cycle by Fourier analysis.[8] The levels of the most important components are shown in Fig. 5.9 and discussed in the next section.

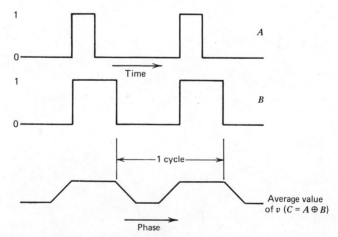

Figure 5.7   Exclusive-OR phase-detector characteristic for non-square waves.

8 Westman, p. 42-12.

5.1.4 Flip-Flop

A flip-flop can be used as a phase detector. This is illustrated in Fig. 5.8. The signals at A and B consist of narrow pulses and are connected to the set and reset inputs of a set-rest flip-flop. The Q output has an average value that is proportional to the phase at A relative to B. A sawtooth characteristic of voltage versus phase is thus created. If either input has finite width, the phase-detector characteristic has a flat spot of corresponding width at a corner. If both inputs have finite width, the flat spot, or the sum of two flat spots, will be as wide as the wider input. One signal may be of arbitrary width without causing such a flat spot if it is connected to the clock input of an edge-triggered flip-flop. One edge of that waveform then defines the switching time. The other input is still connected to a set or reset input and must be narrow if the flat spot is to be minimized.

The chief advantage of the use of a flip-flop is simplicity, a digital I.C. interfacing with other digitial I.C.'s. In the middle of the 360° linear phase range, the output component at the fundamental frequency of the input

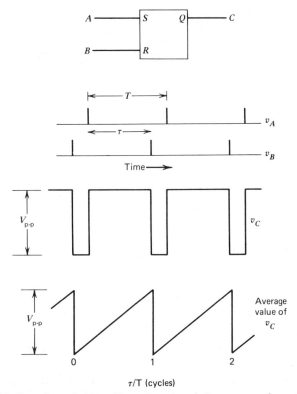

Figure 5.8 Flip-flop phase detector. The pulses A and B cause waveform C, which has an average value that is a function of phase.

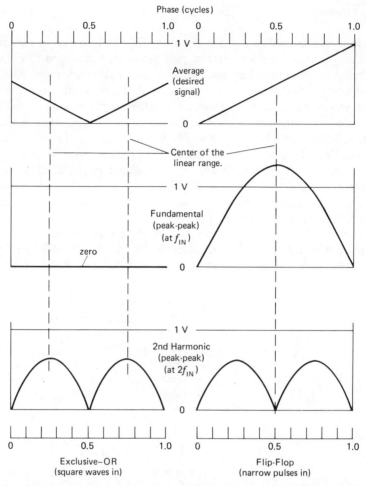

Figure 5.9 Comparison of frequency content of output from exclusive-OR and flip-flop phase detectors with the same $K_\phi = 1$ V/cycle and with given input waveforms.

signals has a peak-to-peak magnitude equal to 1.27 times the peak-to-peak range of the output characteristic, that is, V_{p-p} in Fig. 5.8. This is similar to the exclusive-OR phase detector, but here there is signifcant energy at the input frequency, whereas in the exclusive-OR phase detector, it was at twice that frequency, and thus less troublesome. The (low-) frequency content of the outputs of these two types, each with the same gain, may be compared in Fig. 5.9. Note particularly the magnitudes of the undesired components near the center of the linear range, where operation is more likely to occur. While the flip-flop is disadvantageous relative to the exclusive-OR from the standpoint of undesired signals, it does give twice the range for a given gain, as can be seen from Fig. 5.9.

EXAMPLE 5.1

Problem   A phase-locked synthesizer has the following parameters:

$F_{OUT} = 100$ MHz,
$F_{REF} = 100$ kHz,
$K_v = 1$ MHz/V.

The phase detector is a flip-flop operating at 50% duty cycle between 0 and 5 V. The loop filter is a low-pass filter followed by a dc amplifer.

(a)   What must be the attenuation of the loop filter at 100 kHz to allow only -40-dB sidebands?

(b)   What does this imply about the relationship between f_p and K_{LF}?

(c)   If the phase margin is to be 45°, what will be the value of K_{LF}, f_p, and ω_0?

(d)   What is the hold-in range of this synthesizer?

(e)   If the filter capacitor were removed, (neglecting the resulting FM), how long would it take to get from a frequency output of 99 MHz to within 100 Hz of the final value of 100 MHz?

Solution   We will proceed without reference to a graph and consider the graphical representation later in Example 5.2.

(a)   For -40-dB sidebands at $F_{OUT} \pm 100$ kHz, the peak frequency deviation is given by

$$\Delta f = f_m m,$$

where m is the modulation index, equal to twice the sideband-to-carrier ratio:

$$m = 2 \times 10^{-40/20} = 0.02,$$

Combining these at a modulation frequency of 100 kHz gives a peak frequency deviation of

$$\Delta f = 100 \text{ kHz} \times .02 = 2 \text{ kHz}.$$

At 1 MHz/V, the corresponding VCO input is 0.002 V peak. The amplitude of the fundamental component from a 50%-duty-factor flip-flop phase detector is 1.27 times its mean-to-peak output. The peak value from a 0–5 V phase detector is therefore

$$V_p = 1.27 \times \frac{5}{2} \text{ V} = 3.18 \text{ V}.$$

The required attention is

$$\frac{3.18 \text{ } V}{0.002 \text{ } V} = 1590, \text{ or 64 dB}.$$

(b)   Assuming that 100 kHz is well beyond the filter-corner frequency, the filter gain is given by

$$G_{LF}(100 \text{ kHz}) \approx K_{LF} \frac{f_p}{10^5 \text{ Hz}} = \frac{1}{1590}.$$

This gives

$$K_{LF} f_p = 62.9 \text{ Hz}.$$

(c)   Considering the tangential approximation to the open-loop gain, 45° phase shift will occur at the filter corner frequency, so the loop gain should be one there for 45° margin. This requires

$$\frac{K_\varphi K_v}{N} K_{LF} = \omega_p.$$

The divide ratio is f_{OUT}/f_{REF} or 1000. K_φ is 5 V/cycle and K_v was given as 10^6 Hz/V. Thus we have

$$2\pi f_p = 5 \times 10^3 K_{LF} \sec^{-1}.$$

Combining this with the equation obtained in (b) above, we obtain

$$f_p = \left(\frac{5000 \times 62.9}{2\pi \sec^2} \right)^{1/2} = 224 \text{ Hz}.$$

Using either equation again, we obtain

$$K_{LF} = 0.28.$$

Because the filter corner is at the unity-gain point of the -6-dB/octave region of the open-loop gain curve, we have

$$\omega_0 = \omega_p = 2\pi(224) = 1407.$$

(d) The hold-in range is the phase-detector range times the gain to f_{OUT}, or

$$5 \text{ V}(10^6 \text{ Hz/V})(0.28) = 1.4 \text{ MHz, (peak to peak)}.$$

(e) The response is that of a low-pass filter, as shown in Fig. 3.7c, with filter-corner frequency at ω_0:

$$e^{-t/\tau} = 100 \text{ Hz/1 MHz},$$

$$e^{-\omega_0 t} = 10^{-4} = e^{-2.3(4)}$$

$$t = \frac{(2.3)4}{\omega_0} = \frac{9.2}{1407 \sec^{-1}} = 6.5 \text{ ms}.$$

5.1.5 Sample-and-Hold Phase Detector[9, 10]

In the sample-and-hold phase detector, a ramp is created which is proportional to the time from each reference pulse and thus to the phase of the reference signal. At each occurence of the sample pulse, the value of this ramp is transferred to the hold circuit where it is held until the next sample pulse. This held signal is the output, and is proportional to the phase difference between the sample pulse train, which is synched to the VCO through the frequency divider, and the reference pulse train, which is synched to the reference signal. The waveforms are illustrated in Fig. 5.10 for a case where the sample pulse train is at a slightly higher frequency than the reference pulse train.

The concept of the sample-and-hold circuit is illustrated in Fig. 5.11. The current source produces a voltage ramp across the ramp capacitor, but each

9 Kroupa, pp. 184–187.
10 Manassewitsch, pp. 401–407.

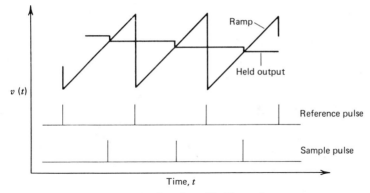

Figure 5.10 Sample-and-hold waveforms.

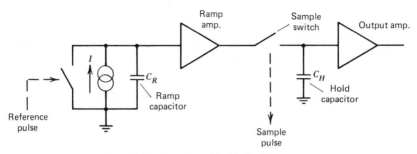

Figure 5.11 Sample-and-hold circuit concept.

reference pulse causes the ramp capacitor to discharge. Thus a series of ramps occur at the reference frequency, producing a sawtooth waveform. The ramp amplifier provides isolation, preventing the ramp capacitor from being discharged by the sample switch. When the sample switch closes, during the sample pulse, the ramp voltage is transferred to the hold capacitor. When the sample switch opens again, the hold capacitor maintains its charge until the next sample. The output amplifier isolates the hold capacitor to prevent its discharge by the driven circuitry.

One possible practical circuit is shown in Fig. 5.12*a* with waveforms in Fig. 5.12*b*. An integrated current source (CS$_1$), consisting of a FET with its gate and source connected internally, supplies current to C_R. C_R is discharged through Q_1 and R_1 during the reference pulse. Q_2 and Q_3 isolate C_R to reduce leakage.

U_1 is an integrated FET switch, activated by the sample pulse. In addition to charging C_H during the sample pulse, U_1 stops the charging of C_R during the time C_H is being charged. If this is not done, the steady-state output will contain transients, as shown in Fig. 5.13. Some circuits eliminate the transients by using two sample switches in series, the second one delayed so as not to close until the first one has opened. Thus, the second switch is closed only

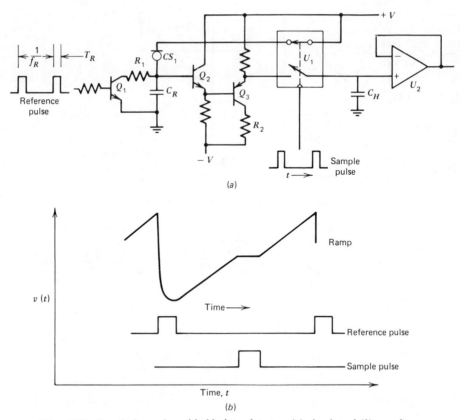

Figure 5.12 Practical sample-and-hold phase detector: (*a*) circuit and (*b*) waveforms.

during the flat part of the waveform in Fig. 5.13, and the transients do not appear at the output. Unfortunately, this introduces additional delay which may be harmful to loop stability.

U_2 is an op-amp connected as a voltage follower.

Ideally, sample-and-hold phase detectors produce only dc under steady-state conditions. In practice, they can be built with outputs at the reference frequency and its lower harmonics of the order of 100 μV rms (for units with a 5-V output range and reference frequencies in the range of 4–60 kHz). This level is about 96 dB lower than the significant output components of the phase detectors discussed above, relative to their output ranges.

One cause of these undesired outputs is stray coupling of the digital waveforms, particularly coupling of the sample pulse to the output. This can

Figure 5.13 Steady-state output waveform when ramp not stopped.

be reduced by coupling a complementary waveform to the output.[11] A second cause is leakage of the hold-capacitor charge through the capacitor or switch or into the output amplifier. The sample switch must then recharge C_H at each sample pulse, producing a sawtooth waveform at the output. Leakage of the charge on C_R will cause droop during the sample pulse and can produce wave shapes like that in Fig. 5.13. In addition, high-dielectric-constant ceramic capacitors tend to produce a voltage change after the current stops flowing.[12] Glass or low-dielectric-constant ceramic capacitors are preferred for the ramp capacitor C_R.

Note that the ramp in Fig. 5.12 has a linear range less than 360° because of the finite pulse widths. Figure 5.14a shows a ramp which has not been stopped for a sample pulse and Fig. 5.14b shows the sampled output voltage versus the time from the beginning of the reference pulse, assuming that the ramp is prevented from rising during the sample pulse. If the two pulses start simultaneously, the ramp may not have reached its minimum by the end of the sample pulse, so a nonzero voltage will be held when the sample switch opens. At greater delays, the output approaches zero. A higher output value does not occur until the beginning of the sample pulse occurs after the end of the reference pulse, since we are assuming the ramp stops during the sample pulse. The output stops rising when the end of the sample pulse occurs at the end of the ramp because, at greater delays, the sample switch is open after the ramp discharge has begun. The usable output range is given by

$$V_R'' = V_R\left(1 - \frac{T_{RW} + T_{SW}}{T_{REF}}\right), \tag{5.10}$$

where

$$V_R = T_{REF}\frac{dv}{dt}. \tag{5.11}$$

The above relationships will be somewhat different if the ramp is not stopped during a sample, as when two sequential sample switches are employed.

The sample switch, and the source driving it, have finite resistances resulting in finite charging time for the hold capacitor. If this resistance is R_S, the voltage change on the hold capacitor is

$$\Delta V(T_{SW}) = \Delta V(\infty)\left[1 - \exp\left(-\frac{T_{sw}}{R_S C_H}\right)\right] \tag{5.12}$$

where $\Delta V(\infty)$ is the difference between the source voltage and the held voltage at the start of the sample pulse. The sampling efficiency is defined as

$$\eta_s \triangleq \frac{\Delta V(T_{SW})}{\Delta V(\infty)} \tag{5.13}$$

$$= \left[1 - \exp\left(-\frac{T_{SW}}{R_S C_H}\right)\right]. \tag{5.14}$$

11 Sherwin.
12 Buchanan.

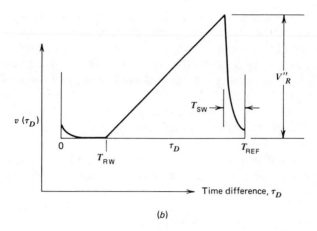

(b)

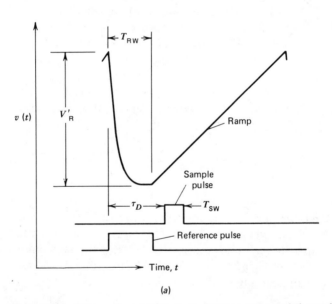

(a)

Figure 5.14 Characteristics of sample-and-hold phase detector: (a) sampled waveform (ramp not stopped during sample) and (b) output versus τ_D (ramp stopped during sample).

This approaches unity as the charging time constant becomes much less than the sample pulse width.

In addition to finite efficiency, real sample-and-hold phase detectors can have limited step sizes due to limited current through the sample switch. Thus it may not be possible to step from one end of the output-voltage range to the other. In addition, the ramp amplifier may be cut off if the step is too large. The size of the step which will cause this depends on the step direction and the output level. A resistor R_2 is shown in the collector of Q_3 in Fig. 5.12a. It is there to protect Q_3 from excessive current surges during sampling, but it also can cause Q_3 to saturate.

Just as, in Fig. 5.2, a high-frequency component at f_{REF} is translated to a low frequency by the sampling process, so too can high frequenices of noise, that precedes the sample switch in the phase detector, be translated to lower frequencies. Thus, the noise components at many higher frequencies can be superimposed upon each other at a frequency which is low enough to pass through the loop. Therefore, even the high-frequency noise in the ramp amplifier can be important; an amplifier with a broad, high-level noise floor can introduce significant noise into the loop.

EXAMPLE 5.2   To illustrate the improvements that can be given by the sample-and-hold phase detector, we give an example similar to the previous example, but with the flip-flop replaced with a sample-and-hold phase detector.

Problem

(a) Do Example 5.1 except assume a sample-and-hold phase detector, with an output range of 5 V for 360° and 100 μV rms output at the sample frequency, and require -60-dB sidebands. Do not consider phase shift due to sampling.

(b) How are the hold-in range and speed affected if (a) is repeated but with a 10-kHz reference frequency?

(c) Consider the effects of sampling phase shift.

Solution

(a) -60-dB sidebands imply $\frac{1}{2}m = 10^{-3}$. Therefore,

$$\frac{\Delta f}{10^5 \text{ Hz}} = 2 \times 10^{-3}, \Delta f = 200 \text{ Hz}.$$

The corresponding peak voltage at the VCO input is

$$V_p = \frac{\Delta f}{10^6 \text{ Hz/V}} = 2 \times 10^{-4} \text{ V}.$$

The filter gain at 100 kHz must not exceed

$$G_{LF}(100 \text{ kHz}) = \frac{(V_{VCO IN})_{PEAK}}{(V_{PD OUT})_{PEAK}} = \frac{2 \times 10^{-4}}{\sqrt{2} \times 10^{-4}} = \sqrt{2},$$

or 3-dB gain.

Using this value, the loop gain at 100 kHz is computed to be

$$GH(100 \text{ kHz}) = \frac{(5 \text{ } V/\text{cycle})\sqrt{2} \text{ } (10^6 \text{ Hz/V})}{(2\pi \times 10^5)(10^3)}$$

$$= 1.125 \times 10^{-2},$$

or -39 dB. This is illustrated by the solid curves of Fig. 5.15. The loop gain reaches unity at 10.6 kHz, which can be shown graphically or can be calculated by determining the change in frequency corresponding to a 39-dB

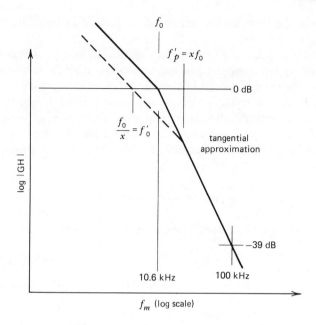

Figure 5.15   Gain plot for Example 5.2.

change on a -40-dB/decade slope:

$$f_0 = 100 \text{ kHz} \times 10^{-39/40} = 10.6 \text{ kHz.}$$

For 45° phase margin, this also equals f_p.

The value of K_{LF} can be determined by the following method, or by graphically extending the curve to $\omega = 1$ and dividing the gain there by all factors other than K_{LF}:

$$\omega_0 = 2\pi f_0 = 6.67 \times 10^4,$$

$$K_{LF} = \frac{6.67 \times 10^4 (1000)}{(5 \text{ V/cycle})(10^6 \text{Hz/V})} = 13.32.$$

The hold-in range is the phase-detector range multiplied by the dc forward gain, or

$$5 \text{ V}(10^6 \text{ Hz/V})(13.32) = 66.6 \text{ MHz peak-to-peak.}$$

The pull-in time, as defined in Example 5.1 (no filter capacitator), is

$$t = \frac{4(2.3)}{6.67 \times 10^4} = 138 \ \mu\text{sec.}$$

Thus a change of phase-detector type has allowed significant and simultaneous improvement in the level of the sampling sidebands, the pull-in time, and the hold-in range.

(b)   If the reference frequency is lowered to 10 kHz, the procedure used above shows that the hold-in range will be reduced from 66.6 to 2.1 MHz and the pull-in time will go from 138 μsec up to 4.37 msec, if the -60-dB sidebands are maintained.

(c) Effect of sampling: The approximate phase shifts at f_0 due to sampling can be computed from Eq. (5.9) as 0.4°, 19°, and 6°, respectively, for Examples 5.1, 5.2(a) and 5.2(b). We will now consider how the inclusion of this phase shift modifies the problem by lowering the frequency at 45° phase margin, using the worst case, Example 5.2(a). If the gain curve is moved to the curve shown dashed in Fig. 5.15, the filter corner will increase in frequency by a factor x and the unity gain frequency will decrease by x. The phase shift due to sampling then becomes, at unity gain,

$$\varphi_s = \frac{19°}{x}.$$

The loop-filter corner frequency now is x^2 greater than the unity-loop-gain frequency, so the contribution of phase from the filter at unity loop gain is

$$\varphi_F = \tan^{-1}\left(\frac{1}{x}\right)^2.$$

At $x = 1/0.77$, these factors sum to

$$\varphi_T = 19°(0.77) + \tan^{-1}(0.77^2) = 14.63° + 30.66° = 45.29°.$$

Therefore, the frequency for unity gain at 45° margin becomes

$$f_0' = 0.77(10.6 \text{ kHz}) = 8.2 \text{ kHz}.$$

The filter corner is at

$$f_p' = \frac{10.6 \text{ kHz}}{0.77} = 13.8 \text{ kHz}.$$

The value of K_{LF} is multiplied by 0.77 to 10.26 kHz. These changes are appreciable but do not basically change the comparison between the three cases studied in Examples 5.1 and 5.2.

5.1.6 The Phase-Frequency Detector[13, 14]

The phase-frequency detector acts as a phase detector during lock and provides a frequency-sensitive signal to aid acquisition when the loop is out of lock. It is available in integrated circuit form and consists of logic circuitry plus, in most versions, parts of the charge pump and amplifier.[13, 15, 16] The charge pump is illustrated in Fig. 5.16. Logic circuitry closes S_2 to raise the value of V_3 or closes S_1 to lower it. The resistances convert the open-circuit voltage into a current which charges a capacitor to produce an output voltage. This capacitor may be part of a low-pass structure, in which case the VCO frequency is controlled by the average voltage output, or it may be part of an integrator circuit, in which case the phase-locked loop becomes a type-2 loop. Type-2 loops will be discussed in some detail later. For now it is enough to say that the loop has zero steady-state phase error because any continuous error would cause the integrator output to continually change.

13 Motorola Data Sheet . . . MC4344.   15 Motorola Data Sheet . . . MC12012.
14 Breeze.   16 Fairchild Data Sheet . . . 11C44.

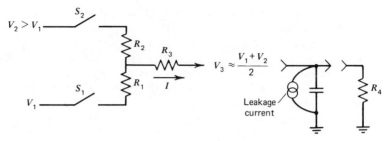

Figure 5.16   Charge pump.

This type of phase detector goes well with the type-2 loop because, if there is no steady-state phase error, and thus S_1 and S_2 never close, no transients will be created. In practice, S_1 or S_2 will close each reference period for a short interval to provide enough charge to the capacitator to compensate for leakage during the period. Still, the shorter the duration of closing the smaller will be the transient.

Operation of the phase-frequency detector is illustrated by the waveforms of Fig. 5.17. There are three output states: positive current, negative current, and no current. The impulses shown will, in practice, correspond to the triggering edges of waveforms. The reference pulse causes the output to change in a positive direction, unless the output is already positive, in which case the pulse has no effect. Similarly, the loop's frequency divider output causes a negative transition unless the output is already negative. The result is illuetrated in the lower waveform for a case where the divider output frequency is higher than the reference frequency. Figure 5.18 is a plot of the average voltage output (through a low-pass filter) versus phase. Note that the phase range is 720° and if, due, let us say, to a minute frequency difference between inputs, the output voltage reaches an extreme, it then returns to zero. This return occurs as two input pulses of the same type occur without an intervening pulse of the other type. However, as long as the frequency error has the same sign, the locus will continue to traverse the same half of the output characteristic. This is of great value in acquisition of lock because the average output voltage in the presence of a frequency error tends to the

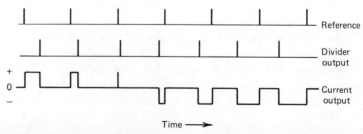

Figure 5.17   Phase-frequency detector waveforms with high VCO frequency.

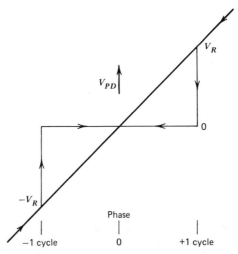

Figure 5.18 Output characteristic of phase-frequency detector. After one pulse of width sufficient to produce an average voltage beyond $\pm V_R$, the operating point passes through zero phase.

correct direction to bring the circuit into lock. This will be further considered later when acquisition is studied.

Of primary concern in this type of phase detector is crossover distortion,[17] changes in gain that can occur near zero phase error. Figure 5.19 shows equal-amplitude current pulses of both polarities with finite delay, rise and fall times. These two pulses would not actually occur during the same period, since they occur for different divider output times. Note that, as the divider output comes closer to the reference, there is a linear change in charge until the pulse (either i_U or i_D) rise and fall transients meet. Then there is a

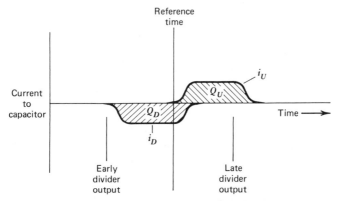

Figure 5.19 Positive and negative current pulses. From Egan and Clark, p. 136. Reprinted with permission from *Electronics Design*, Vol. 26, No. 12, copyright Hayden Publishing Co., Inc., 1978.

17 Egan and Clark.

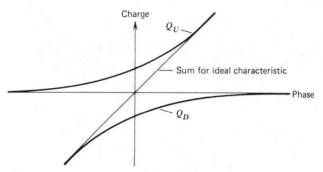

Figure 5.20 Charge versus phase, ideal. From Egan and Clark, p. 136. Reprinted with permission from *Electronics Design*, Vol. 26, No. 12, copyright Hayden Publishing Co., Inc., 1978.

decreasing rate of charge change with time difference between reference and divider output. This is shown in curves marked Q_U and Q_D in Fig. 5.20. Here the delay, rise and fall times of the pulses are properly matched so the total charge is a linear function of phase. (R_1 and R_2 could be only internal resistances, but we will assume that they are much larger than R_3, for simplicity, so we will not have to account for the influence of one switch on the current from the other.) If, however, i_D were delayed more (Fig. 5.21), the slope of the output characteristic would become greater, up to twice the value it has in the noncrossover region. In the other direction, as the i_D pulse terminates sooner, relative to the start of the i_U pulse, a flat region develops at crossover and the gain can actually go to zero. This is illustrated in Fig. 5.22. Such a situation has actually been observed in I.C. phase detectors. Both pulses have been seen to effectively disappear simultaneously in a narrow region near crossover. The result, of operation in a region such as this, is an open loop. As the lock point moves through this region there is no loop gain.

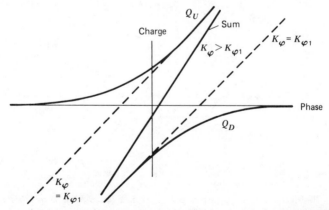

Figure 5.21 Charge versus phase, excess gain at crossover. From Egan and Clark, p. 137. Reprinted with permission from *Electronics Design*, Vol. 26, No. 12, copyright Hayden Publishing Co., Inc., 1978.

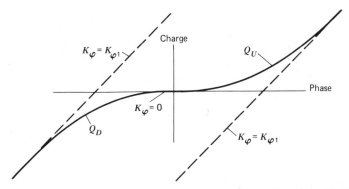

Figure 5.22 Charge versus phase, zero gain at crossover. From Egan and Clark, p. 137. Reprinted with permission from *Electronics Design*, Vol. 26, No. 12, copyright Hayden Publishing Co., Inc., 1978.

Noise would normally rise on the synthesizer output and the frequency can wander about over a restricted range. Moreover, gains of the order of 10 times nominal have been observed near crossover in the same circuits that have dead zones at crossover (see Fig. 5.23). These may be due to internal logic race conditions.[18] Obviously, such high gains can lead to altered loop characteristics, including instability.

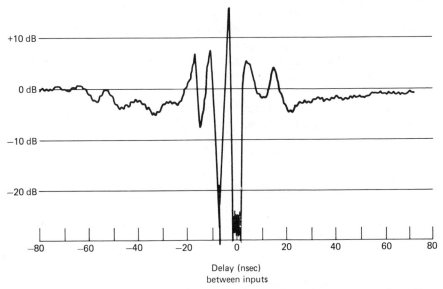

Figure 5.23 Measured gain (K_ϕ) of a phase-frequency detector near crossover. Similar to Egan and Clark, p. 135. Reprinted with permission from *Electronics Design*, Vol. 26, No. 12, copyright Hayden Publishing Co., Inc., 1978.

18 Fairchild Data Sheet . . . 11C44, p. 5.

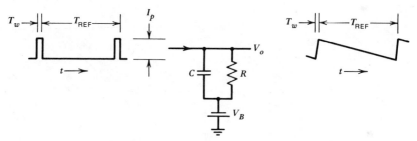

Figure 5.24 Current pulses into a low-pass filter.

There is nothing which can be done following the digital circuitry to eliminate the zero-gain region if there are no pulses to shape. However, its width and the overall gain distortion can be affected by the subsequent circuitry.

A way to reduce deleterious effects of crossover distortion on the loop is to design in enough leakage current to insure operation far enough removed from crossover, under steady-state conditions. This, of course, increases the reference frequency content in the output. The increase will be smaller, however, if the "leakage" current is supplied as a narrow pulse near crossover.[19] This pulse, and the compensating pulse from the charge pump, could, in theory, be made to cancel each other but, even if they do not overlap, their contribution to troublesome low-order multiples of the reference frequency can be reduced by their proximity.

Note, also, that the gain for positive and negative pulses is not necessarily the same. Also, as V_3 in Fig. 5.16 changes, the amplitude of one pulse will increase and that of the other will decrease. Thus, if an RC filter is used, the gain will change somewhat with phase.

If an RC low pass is used, as shown in Fig. 5.24, assuming the filter is low enough in frequency to be effective against f_{REF} and assuming small ripple, the voltage droop during a reference period is

$$\Delta V \approx \frac{(T_{REF} - T_w)(\overline{V}_0 - V_B)}{RC} , \qquad (5.15)$$

where $\overline{V}_0$ is the average output voltage. The nth harmonic of the output waveform has an amplitude of[20]

$$V = \Delta V \frac{T_{REF}^2}{\pi^2 n^2 T_w (T_{REF} - T_w)} \sin\left(n\pi \frac{T_w}{T_{REF}} \right) \qquad (5.16)$$

$$= \left(\overline{V}_0 - V_B \right) \frac{f_p}{F_{REF}} \alpha_n, \qquad (5.17)$$

where f_p is the filter cutoff frequency, V_B is the output voltage when T_w is zero,

19 Breeze, p. 13.
20 Westman, p. 43-13.

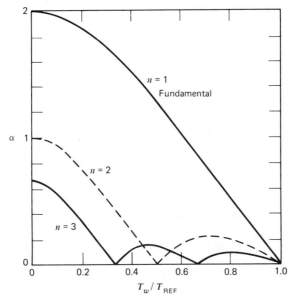

Figure 5.25 α versus T_w/T_{REF}.

and α is shown in Fig. 5.25 for the first three harmonics. From Fig. 5.25 and Eq. (5.17) can be seen that the output voltage at the reference frequency has an effective amplitude, before reduction by the filter, of up to twice the average (dc) output offset $(\overline{V}_o - V_B)$. For small T_w, harmonics are apparently smaller than the fundamental by their harmonic number. This is not a very quiet phase detector when used this way, but that should be no surprise since it produces wide rectangular pulses, when $\overline{V}_o$ is not close to V_B.

Figure 5.26 shows a more usual type of filter for this phase detector, an integrator with a zero, which is required for loop stability. As has been noted,

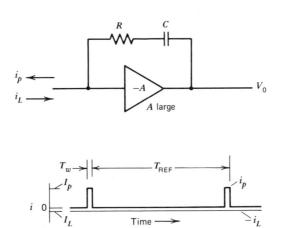

Figure 5.26 Current pulses into an integrator with a zero.

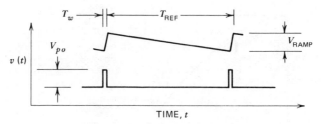

Figure 5.27　Components of output.

the current pulse into this filter (part of a type-2 loop) will be narrow at steady state. The two components of the waveform produced at the filter output are shown in Fig. 5.27, one due to the drop across C and one due to the drop across R. The voltage across C has a peak-to-peak magnitude of

$$V_{\text{RAMP}} \approx \frac{I_L T_{\text{REF}}}{C} \tag{5.18}$$

where I_L is the leakage current. The amplitude of the nth harmonic of the capacitor voltage is[21]

$$V_n' = V_{\text{RAMP}} \frac{1}{\pi n} \tag{5.19}$$

$$\approx \frac{1}{\pi n} \frac{I_L T_{\text{REF}}}{C} . \tag{5.20}$$

The amplitude of the voltage across the resistor is

$$V_{po} = I_p R \tag{5.21}$$

and the nth-harmonic amplitude is

$$V_n'' = 2I_p R \frac{T_w}{T_{\text{REF}}} \frac{\sin(n\pi T_w / T_{\text{REF}})}{n\pi T_w / T_{\text{REF}}} . \tag{5.22}$$

If T_w is much smaller than the period of the nth harmonic, this is

$$V_n'' \approx 2I_p R \frac{T_w}{T_{\text{REF}}} . \tag{5.23}$$

Equating the charge acquired during the pulse, $I_p T_w$, to that lost through leakage, $I_L T_{\text{REF}}$, this becomes

$$V_n'' = 2I_L R. \tag{5.24}$$

The ratio of the amplitudes of the two components of the nth harmonic is

$$\frac{V_n''}{V_n'} = \frac{2\pi n R C}{T_{\text{REF}}} \tag{5.25}$$

$$= n \frac{f_{\text{REF}}}{f_z} , \tag{5.26}$$

21　Westman, p. 42-12.

where f_z is the frequency of the filter zero, $1/(2\pi RC)$. For stability, f_z is generally much lower than f_{REF}, so V_n'' predominates. Therefore, the amplitude of the fundamental and lower harmonics resulting from a leakage current I_L into an integrator with a resistor R used to produce a zero is

$$V_n \approx 2I_L R. \tag{5.27}$$

EXAMPLE 5.3

Problem A charge pump drives an "integrator" which has a 0.1-μF capacitor and a 4000-Ω resistor in series in the feedback path. The total leakage current into the capactor is 2 μA. The "integrator" output is connected to a VCO with a 10-MHz/V sensitivity. What is the steady-state duty cycle of the charge pump if it generates 0.5-mA current pulses? What is the peak deviation of the VCO output component at a modulation frequency equal to twice the reference frequency?

Solution The duty cycle will be the ratio of leakage current to pulse current:

$$\text{Duty cycle} = \frac{2 \times 10^{-6}}{5 \times 10^{-4}}$$
$$= 0.004.$$

Since this is a small part of the second-harmonic period, the approximation in Eq. (5.24) holds and the peak amplitude of the second harmonic is

$$V_2 = 2(2 \times 10^{-6} \text{ A})(4000 \ \Omega)$$
$$= 16 \text{ mV}.$$

We multiply this by the tuning sensitivity to obtain the peak deviation at the second harmonic (and the fundamental):

$$f = 16 \times 10^{-3} \times (10^7) \text{ Hz}$$
$$= 160 \text{ kHz}.$$

5.2 EFFECTS OF SAMPLER ON LOOP TRANSFER FUNCTION

Initially, we represented the phase-detection process only by a frequency comparison, or subtraction, followed by integration of the frequency difference to give phase difference. When a sample-and-hold phase detector is used, it may become necessary to further refine this approximation to include the effects of the sample-and-hold process. This is not so much because sampling only occurs with this type of phase detector—we have shown (Section 4.5) that the digital frequency divider only transmits phase information at discrete times—but more because the lower content of components at the reference frequency allows less filtering and wider bandwidths, so the effects of sampling at the loop corner can become significant in determining stability and response. The charge-pump phase detector also can be considered to produce a sampled output approximately at the time of the divider transition which triggers the current on or off.

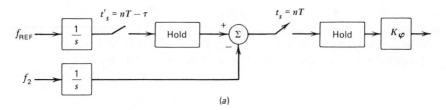

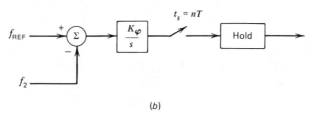

Figure 5.28 Phase sampling and summing. The differing sampling times in (a) are taken as equal in the approximation (b).

We have noted, in Section 3.5, that the sampled phase is determined by the phase of the reference signal at the beginning of the ramp, a time earlier than the sample time. This situation is presented in Fig. 5.28a, where the reference phase is sampled at the beginning of the ramp. The phase of the divider output is subtracted from this and the difference phase is sampled at the end of the divider cycle. We then approximate the time difference τ between these two samples to be zero, in order to achieve the simpler diagram of Fig. 5.28b. This makes no difference in the characteristic equation, since the loop has not changed, and the delay of the reference sample by part of a reference period will not usually be significant. For simplicity, we also approximate the sample frequency as being equal to ths reference frequency, rather than the actual divider output, much as in Section 4.5.

The effect of sampling, and of the hold circuit, upon the sampled spectrum was described in Section 5.1.2. In this section we will consider only the fundamental modulation frequency which is passed by the sampling process, ignoring higher frequencies created by that process.[22] Our justification for doing this is based on the fact that the loop filter attenuates higher frequencies more severly and the loop gain falls 6 dB/octave, even in the absence of a loop filter, due to the phase-detection process.

Figure 5.29 shows a loop with a sampler after the phase detector, followed by a (zero-order) data hold. The output of the hold circuit equals the phase difference between feedback and reference signals, sampled once each reference period and held until the next. Our approximation, which ignores all but the original, before-sampling, modulation, as indicated by Fig. 5.30, is shown

22 Ragazzini, pp. 123–129.

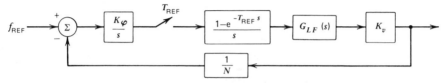

Figure 5.29 Loop with sample-and-hold.

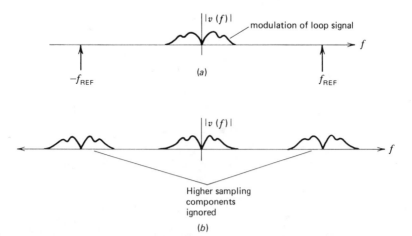

Figure 5.30 Spectral components created by sampling process are ignored: (*a*) original spectrum and (*b*) sampled spectrum.

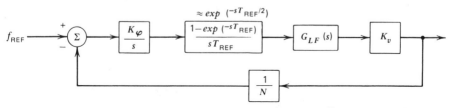

Figure 5.31 Loop where higher-frequency components due to sampling are not considered.

in Fig. 5.31. Figure 5.32 shows the magnitude and phase of the zero-order hold transfer function divided by T_{REF}. Note that the phase reaches 45° before the magnitude drops 1 dB and 20° before a 0.14-dB magnitude reduction. That is why the effect on phase has more impact on the open-loop transfer function plot of Fig. 5.33 and why the sample-and-hold process can often be approximated as a pure delay of $\frac{1}{2} T_{REF}$ in the region where higher frequencies generated by the sampler can be ignored. Figure 5.33 represents a loop where a low-pass filter cuts off at $\frac{1}{8}$ of the sample (reference) frequency. Whereas, without considering the hold function instability cannot occur because 180° excess phase cannot be reached, with the hold function

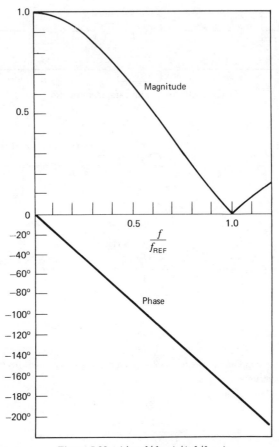

Figure 5.32 $(\sin \pi f/f_{REF})/(\pi f/f_{REF})$.

zero phase margin can occur within an octave of the low-pass corner. The open-loop unity-gain frequency must be lower than this for stability.

A second parameter affecting open-loop gain is sampling efficiency, defined by Eq. (5.13). With finite sampling efficiency η, the hold transfer function becomes[23]

$$G_H(s) = \left[\frac{1 - \exp(-T_{REF}s)}{s} \right]\left[\frac{\eta}{1 - (1 - \eta)\exp(-T_{REF}s)} \right]. \quad (5.28)$$

We will call the second factor in this equation the efficiency factor. It is plotted in Figs. 5.34 and 5.35.

Another modification of the open-loop phase plot that may be warranted is the introduction of a delay (phase shift equal to ω times the delay) equal to

23 Alonzo.

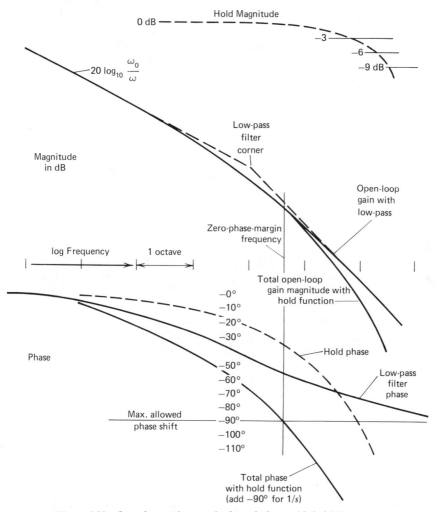

Figure 5.33 Open-loop gain magnitude and phase with hold function.

roughly the time for the voltage on the hold capacitor to complete one-half of its change during the sample pulse. In many cases this may not be significant but the effect should be considered, especially if the charge time constant is an appreciable part of the reference period. Another delay that might sometimes be significant is the propagation delay in the frequency divider (Section 4.5). For example, when a sample-and-hold phase detector is used, this is the time from the divider input transition that triggers the sampling to the initiation of the sample; the effective delay due to the sampling process is included in our representation of the sample-and-hold phase detector and should not be included in the divider.

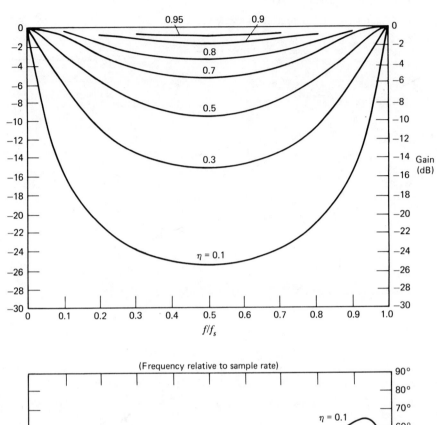

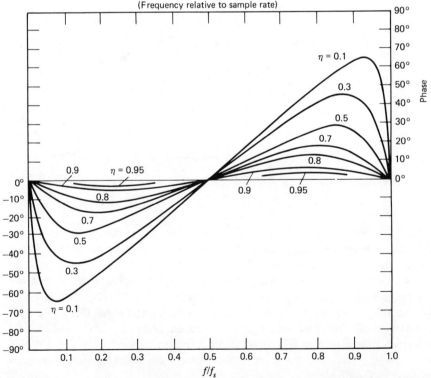

Figure 5.34 Attenuation and phase shift due to efficiency factor.

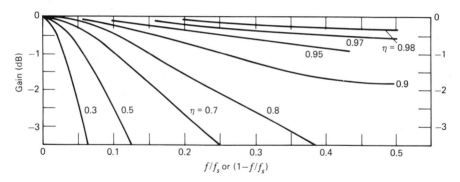

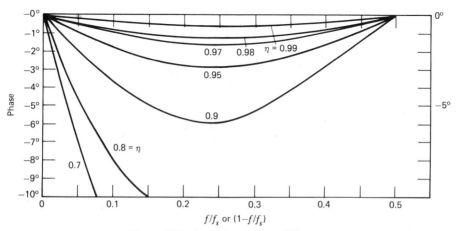

Figure 5.35 Expanded portion of Fig. 5.34.

EXAMPLE 5.4

Problem Draw an approximate open-loop gain and phase plot for a loop with the following parameters:

$K_\varphi = 1$ V/cycle
$K_{LF} = 10$,
$f_p = 10$ kHz,
$K_v = 10$ MHz/V,
$f_{REF} = 40$ kHz,
$N = 1$.

Draw the plot from $\omega = 1$ to $\omega = 10^5$.

(a) Draw the gain and phase curves ignoring the sampling.

(b) Add the effect of ideal sampling, using the simplest approximation.

(c) Add the effect of 80% sampling efficiency.

(d) Assume that the response delay during a sample pulse is equivalent to a 4-μsec time delay and show this effect.

(e) What is the divide number for a 30° phase margin with all of the above effects accounted for?

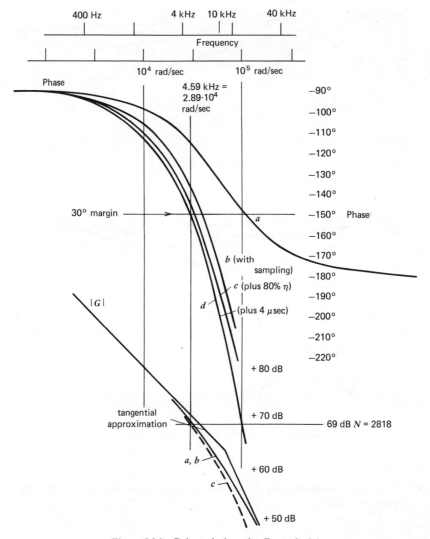

Figure 5.36 Gain and phase for Example 5.4.

Solution

$$\omega_0 = \frac{K_\varphi K_{\mathrm{LF}} K_v}{N} = 10^8 \rightarrow 160 \text{ dB}.$$

See Fig. 5.36. The gain at $\omega = 0$ is 160 dB and drops at 20 dB/decade to 80 dB at $\omega = 10^4$. The slope of the tangential approximation goes to -40 dB/decade at the filter-corner frequency, 10 kHz.

(a) The gain and phase plots are shown as curves a in Fig. 5.36. Only the regions of prime interest are shown.

(b) The simplest approximation for the effect of ideal sampling is a pure time delay of one-half reference period, giving a phase shift of

$$\varphi = 180° \left(\frac{f}{f_{REF}} \right),$$

which is added to the phase shift of curve a. The result is shown in curve b.

(c) The effects of 80% sampling efficiency are taken from Figs. 5.34 and 5.35 and added to curves b in Fig. 5.36. The results are shown in curves c.

(d) A 4-μsec time delay results in a phase shift of

$$\varphi = 4 \, \mu\text{sec} \, (f)(360°),$$

that is,

$$4 \times 10^{-6}(360°)(f \text{ in Hz}).$$

This is added to the previous sum in curve d.

(e) A 30° margin occurs (at $-150°$ open-loop phase shift) at 4.59 kHz, as shown in Fig. 5.36. At this frequency, the gain is seen to be 69 dB when N is 1. To reduce this gain to unity, the divide ratio must be given by

$$20 \log_{10} N = 69.$$

Therefore, N is given by

$$N = 10^{69/20} = 2818.$$

5.3 Z-TRANSFORM REPRESENTATION

If the frequency components created by the sampling process become important, Z transforms can be used to give a more accurate analysis of the loop.[24] This is not usually necessary, but may give some surprising results if the loop bandwidth becomes wide enough relative to the sampling frequency, as we shall see below.

In the absence of a loop filter, the Z transform of the open-loop transfer function for Fig. 5.29 is

$$GH(z) = \frac{\omega_0 T_s}{z - 1}, \tag{5.29}$$

where.

$$z \triangleq e^{sT_s} \tag{5.30}$$

and T_s is the sample period, which we approximate as being equal to T_{REF}. The locus of the closed-loop pole in the Z plane is shown in Fig. 5.37. The root locus is developed from the open-loop singularities according to the same geometric rules as are applied to the S plane for continuous systems, but the region of stability is within the unit circle. In the absence of a filter, the region of instability is entered at a gain of

$$\omega_0 = \frac{2}{T_s}. \tag{5.31}$$

24 Ragazzini.

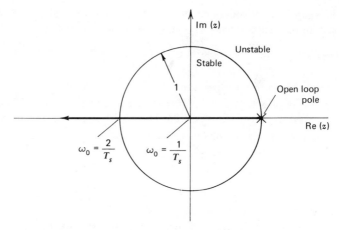

Figure 5.37   Z-plane locus of closed-loop pole for no loop filter.

Let us compare this value to what would be given by the continuous approximation. According to the previous section, an excess phase of 180° would occur when the hold transfer function had a phase shift of 90°. This occurs at

$$\frac{f}{f_s} = \frac{1}{2}. \tag{5.32}$$

At this frequency the hold function has a magnitude of

$$G_H = \frac{\sin\left(\frac{1}{2}\pi\right)}{\frac{1}{2}\pi} = \frac{2}{\pi}. \tag{5.33}$$

Thus, for stability, the remaining loop transfer function ω_0/s must have a magnitude no greater than $\frac{\pi}{2}$ at f equal to $f_s/2$:

$$\frac{\omega_0}{\omega} = \frac{\omega_0}{2\pi(f_s/2)} \leqslant \frac{\pi}{2}, \tag{5.34}$$

$$\omega_0 \leqslant \frac{\pi^2}{2T_s} = \frac{4.9}{T_s}. \tag{5.35}$$

This is 2.5 times more gain than is actually allowed according to the Z-transform analysis, and illustrates the inaccuracy of the approximation, which ignores higher-frequency components created by sampling, as the loop bandwidth becomes appreciable relative to the sampling frequency.

With a low-pass loop filter, with its corner at ω_p, the open-loop transfer

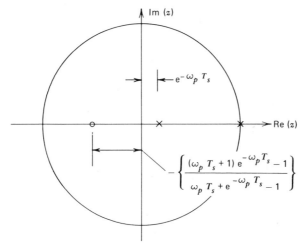

Figure 5.38 Open-loop singularities in the Z plane with low-pass filter.

function is

$$GH(z) = \left[\frac{T_s}{z-1} - \frac{1 - \exp(-\omega_p T_s)}{\omega_p [z - \exp(-\omega_p T_s)]} \right] \omega_0. \tag{5.36}$$

The zero and poles of this function are shown in Fig. 5.38 and the root loci for several filter-corner frequencies are shown in Fig. 5.39. Note that, with a low filter corner frequency, the loop becomes unstable at a gain where the poles are complex but, as the filter frequency increases, a higher gain is required for instability and, eventually, the pole is real at the point of instability. As the filter frequency increases, the open-loop zero approaches an open-loop pole. At infinite filter frequency the two cancel, giving Fig. 5.37.

By means of such analysis, the maximum gain allowed for stability has been plotted as a function of the low-pass pole frequency (A in Fig. 5.40). The plot actually consists of segments of two curves; for $[1/(\omega_p T_s)] \gtrsim 0.27$ the closed-loop poles are imaginary (as in Fig. 5.39a and b) and below this they are real (as in Fig. 5.39c). Also plotted are the maximum gains based on the two continous-system approximations that have been discussed. Note that, for this loop, the simpler approximation (curve C) is accurate to about 10% in allowed gain if the filter corner frequency is less than one-fifth of the sampling frequency while this accuracy may be obtained at up to a third of the sampling frequency if the amplitude of the hold function is included (curve B).

For this loop, the low-pass filter actually improves stability (gain margin), especially if its corner frequency f_p is at about $0.6f_s$, so removal of a low-pass could easily cause instability. This observable fact would not be predicted without the Z-transform analysis.

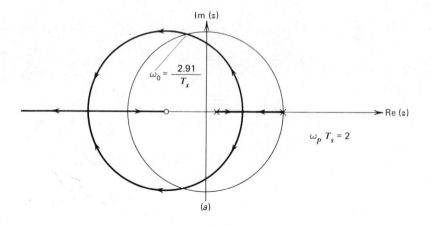

(a)

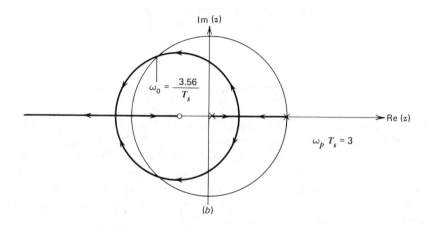

(b)

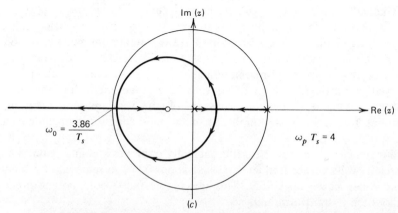

(c)

Figure 5.39 Z-plane locus of closed-loop poles for various values of ω_p.

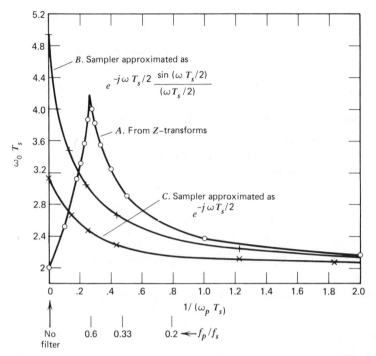

Figure 5.40 Maximum loop gain ω_0 for stable operation versus low-pass corner frequency ω_p, both normalized to sampling period T_s.

5.4 FILTER PLACEMENT

While the placement of the low-pass filter in the forward-gain part of the loop might appear arbitrary, there is some advantage to placing it close to the VCO so it can act on all noise entering before that point. However, it may sometimes be necessary to employ some filtering prior to any amplifier in the forward path to keep it from being saturated by noise spikes from the phase detector.

PROBLEMS

5.1 A phase-locked synthesizer has the following parameters:
$$f_{OUT} = 10 \text{ MHz},$$
$$f_{REF} = 100 \text{ kHz},$$
$$K_v = 100 \text{ kHz/V}.$$
The phase detector is a flip-flop operating at 28% duty cycle between 0.3 and 2.8 V. The loop filter is a low-pass filter followed by a dc amplifier. Use the tangential approximation of the Bode gain plot.

(a) What must be the attenuation of the loop filter at 100 kHz to allow only -60-dB sidebands?

(b) What does this imply about the relationship of f_p and K_{LF} (express in an equation).

(c) If the phase margin is to be 45°, what will be the values of K_{LF}, f_p, and ω_0.

(d) What is the hold-in range of this synthesizer?

(e) If the filter capacitor were removed (neglecting the resulting FM), how long would it take to get from a frequency output of 9 MHz to within 10 Hz of the final value of 10 MHz?

5.2 (a) Do Problem 5.1 except assume a sample-and-hold phase detector with an output range of 5 V for 320° and 300 μV rms output at the sample frequency. Neglect the effect of sampling on the loop transfer function.

(b) Repeat (a) except with a 1-kHz reference frequency.

5.3 Prove the answer given in Example 5.2(b). [i.e., work part (b)].

5.4 Draw an approximate open-loop gain plot for a loop with the following parameters:

$K_\varphi = 1$ V/rad,

$K_{LF} = 1$,

$f_p = 2$ kHz,

$K_v = 10$ MHz/V,

$f_{REF} = 10$ kHz,

$N = 1$.

Draw the plot from $\omega = 1$ to $\omega = 10^5$.

(a) Draw the gain and phase curves ignoring sampling.

(b) Add the effect of ideal sampling, using the simplest approximation.

(c) Add the effect of 70% sampling efficiency.

(d) Assume that the response delay during a sample pulse is equivalent to a 25-μsec time delay and show this effect.

(e) What is the divide number for a 45° phase margin with all of the above effects accounted for?

5.5 Show that Eq. (5.20) can be used to represent the effect of leakage from the hold capacitor in a sample-and-hold phase detector where the sample pulse is narrow compared to T_{REF}. Compute the magnitude of the second-harmonic component due to this effect if leakage is 1 μA from a 0.01-μF capacitor and the reference frequency is 1 kHz.

5.6   A charge pump drives an "integrator" which has a 1 μF capacitor and a 1-kΩ resistor in series in the feedback path. The total leakage current into the capacitor is 10 μA.

(a)   What is the amplitude of the voltage across the resistor at the reference frequency?

(b)   What is the steady-state duty cycle of the current pulse from the charge pump if its magnitude is 1 mA?

(c)   If the "integrator" output is connected to a VCO with a 1-MHz/V sensitivity, what is the peak frequency deviation at the reference frequency?

6
Frequency Dividers

The most apparent feature which differentiates the typical phase-locked synthesizer from other phase-locked loops is the frequency divider. It requires a significant part of the overall design effort in many synthesizers and offers an opportunity for creative design in minimizing cost, space, or power consumption and attaining the necessary operating speed (frequency).

We have noted* that a number of different means exist for frequency division, but the most generally useful in synthesis is the digital divider. In this chapter, we will concentrate on digital dividers.[1, 2] We will assume, in the following, that binary one is represented by a higher voltage than binary zero.

6.1 FLIP-FLOPS

Dividers are composed basically of gates and flip-flops. There are three main types of flip-flops, the set-reset (S-R), D and J-K flip-flops.

The S-R flip-flop is shown in Figure 6.1 along with a mechanization consisting of NAND gates. As long as both inputs are ones, the output remains unchanged. Bringing one input (A or B) to zero causes a one at either S or R. A one at S sets the flip-flop, that is, causes Q to be one. A one at R resets it, causing Q to be zero. When that input returns to one, the last state of the flip-flop remains. The S-R flip-flop is basically a memory element.

The D flip-flop, on the other hand, is basically a shift-register element. On the effective edge of the clock waveform (i.e., the specified or proper edge to cause a reaction), the Q output becomes equal to the D input. Figure 6.2 shows a D flip-flop and external connections which make it into a binary divider. With these connections, the D input is always of opposite state to the

---

* See Section 1.4.3.

1 Kroupa, pp. 66–79.
2 Lee.

Figure 6.1 *S-R* flip-flop.

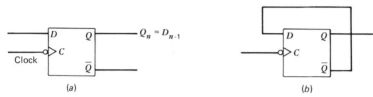

Figure 6.2 *D* flip-flop.

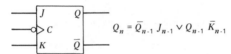

$$Q_n = \overline{Q}_{n-1} J_{n-1} \vee Q_{n-1} \overline{K}_{n-1}$$

Figure 6.3 *J-K* flip-flop.

Q output, so the Q output changes state at each effective clock transition. The symbol shown is for a flip-flop which responds to negative-going edges of the clock waveform.

Probably the most commonly used divider element is the J-K flip-flop, shown in Fig. 6.3. On the (proper) clock edge, if $\overline{Q}$ is zero, $\overline{Q}$ will become one if K is one; otherwise, there will be no change. If Q is zero, Q will become one if J is one; otherwise, no change occurs. In other words, the next state is determined by the input (J or K) corresponding to the output (Q or $\overline{Q}$ respectively) which is zero. If both J and K are one, the circuit will divide the clock frequency by 2. If both are zero, it will not change states.

The highest-frequency digital dividers operate at input frequencies in excess of 1 GHz but function only as fixed dividers with no logic inputs (J, K or D).[3]

Toggle frequencies as high as 500 MHz are achievable with emitter-coupled logic (ECL) D flip-flops.[4] Power consumption is high, 0.2 W per flip-flop, and interconnections must be made using microwave techniques because rise and fall times are only slightly more than 1 nsec.[5] Slower ECL circuits are also available with correspondingly less stringent interconnections and lower power consunmption. The higher logic level is about 0.9 V below the positive supply and the logic swing is only about 0.85 V.

3 For example, Plessey SP8750-8752.
4 For example, Fairchild, 11C06 & Motoroia, MC1690.
5 Blood.

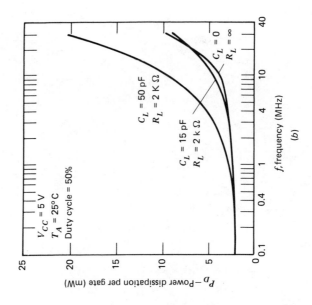

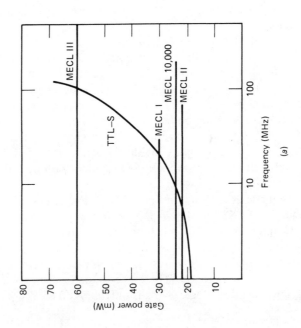

140

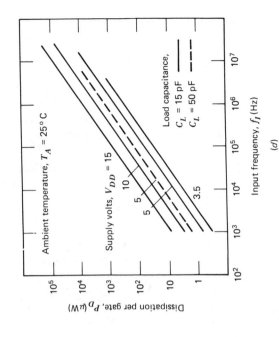

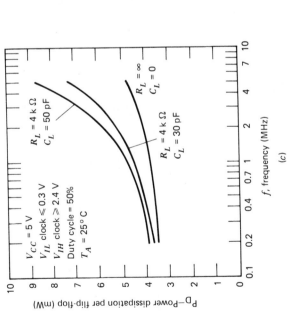

Figure 6.4 Power dissipation versus frequency for various logic types: (*a*) MECL and TTL-S (Schottky) gates; (*b*) low power TTL-S gates; (*c*) low-power TTL flip-flops; (*d*) CMOS gates. [(*a*) Courtesy of Motorola, Inc. Semiconductor Group. (*b*) and (*c*) Courtesy of Texas Instruments Incorporated. (*d*) Courtesy RCA Solid State.]

141

Transistor-transistor logic (TTL) provides logic swings of about 3 V with the low level near ground. Using Schottky-clamped versions, (TTL-S), toggle frequencies slightly above 100 MHz have been available with 75-mW flip-flops or 45 MHz with 10-mW units. These are the power levels at low frequencies, however, and 2.5 or more times this power may be required by these dividers when they operate at the frequencies given above. This is apparent in Fig. 6.4a where the freqnency dependent power dissipation of a TTL-S gate is compared to the frequency-independent dissipation of four types of ECL gates and in Fig. 6.4b where the strong influence of frequency on the dissipation in a low-power TTL-S gate is illustrated. The power consumption of standard (non-Schottky-clamped) TTL logic is less dependent on operating frequency, as can be seen in Fig. 6.4c. Note, however, that the dissipation of the low-power TTL flip-flop becomes highly dependent on frequency when it is loaded by large capacitances.

Beginning in 1979, TTL-S circuits based on more advanced technology (Fairchild 54/74F and Texas Instruments 54/74AS and ALS) began to appear. These circuits are capable of performance ranging from about twice the speed (200 MHz) of the fastest earlier Schottky circuits at the same static power dissipation to one-half the dissipation of the previous lowest-power versions with the same or greater speed.

Complementary metal-oxide (CMOS) devices operate slower and with very low power consumption, but their power consumption increases greatly with increasing frequency.[6] These facts are apparent in Fig. 6.4d.

The dissipation and speed characteristics of many logic families are illustrated in Fig. 6.5.

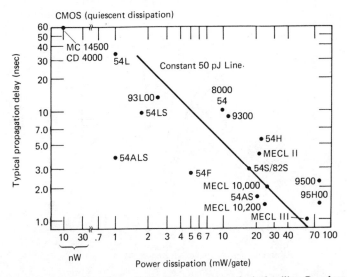

Figure 6.5 Speed versus static power dissipation for various logic families. Based on original, courtesy Motorola Inc. Semiconductor Group.

6 Funk.

6.2 FREQUENCY DIVIDER TYPES

There are two main types of dividers, synchronous and asynchronous. In the former, every flip-flop is triggered by the divider input signal (clock). In the latter, the input signal triggers the first flip-flop, this triggers the second, etc., Synchronous dividers, therefore, achieve a complete transition faster.

Figure 6.6 illustrated an asychronous three-stage divider. A way to obtain an output which is synchronized to the input from this asynchronous divider, at one-eighth of the input frequency, is illustrated in Fig. 6.7. Note how the output of each divider stage lags the preceding stage, but the AND-gate output is only slightly behind the divider input f_{IN}. An AND circuit such as this can help prevent excessive jitter, which would result in phase noise at the synthesizer output.

A synchronous divider is shown in Fig. 6.8. Here the clock triggers each stage, but logic signals to the J and K inputs determine whether the state changes. Thus, all stages that change on a given clock change almost simultaneously. While this makes the last stages respond more rapidly than otherwise, it can cause spikes in a gate output such as in Fig. 6.9. If the stage(s) which is going to zero is slower than the stage(s) going to one, all stages can be simultaneously at one momentarily, causing a spike in the gate output. This "race" problem is illustrated in Fig. 6.9. The spike may be harmful, depending on what the gate is driving. The spike can be prevented by including the input signal f_{IN} among the AND gate inputs, since it will go to zero before the divider outputs change. The asynchronous divider is unlikely to have this problem regardless of what Q ontputs are ANDed, because of the staggered nature of those outputs, as illustrated in Fig. 6.7. If,

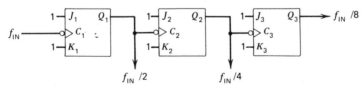

Figure 6.6 Asynchronous divider.

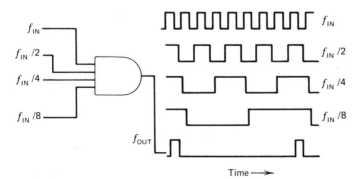

Figure 6.7 Synchronous output from asynchronous divider.

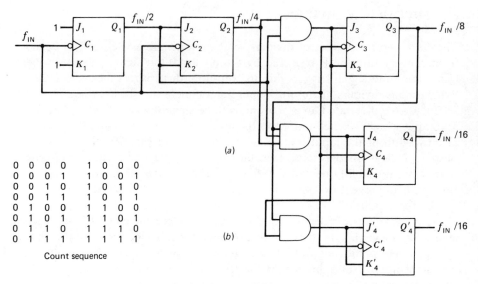

```
0  0  0  0      1  0  0  0
0  0  0  1      1  0  0  1
0  0  1  0      1  0  1  0
0  0  1  1      1  0  1  1
0  1  0  0      1  1  0  0
0  1  0  1      1  1  0  1
0  1  1  0      1  1  1  0
0  1  1  1      1  1  1  1
```

Count sequence

Figure 6.8 Synchronous divider, two versions.

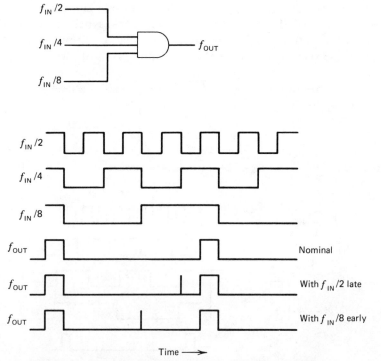

Figure 6.9 Race in synchronous divider producing output spikes.

144

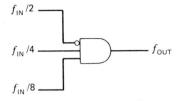

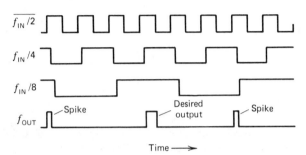

Figure 6.10 Spikes produced when highest frequency input from an asynchronous divider is inverted.

however, the highest frequency input to the AND gate were inverted or, equivalently, taken from the $\overline{Q}$ side of a bistable stage, spikes would be very likely, because the $\overline{Q}$ output would go to one before the succeeding Q outputs went to zero. This is illustrated in Fig. 6.10. However, we are primarily interested in sensing the final number in an increasing count sequence, and it is never necessary to sense a zero for this purpose. At the first occurrence of any set of ones in the (up) count sequence, all other outputs will be zero; since the first occurrence is the only occurrence for the final count, the unsensed numbers are known to be zeros and $\overline{Q}$ outputs need not be used.

EXAMPLE 6.1

Problem In Fig. 6.6 the maximum first-stage delay is 30 nsec, the maximum second-stage delay is 49 nsec, and the maximum third-stage delay is 124 nsec. (a) Delays are equal for positive and negative transitions and the input is a square wave. What is the highest input frequency that can be tolerated without the possibility of spikes in the output of the NAND gate in Fig. 6.7? What is the highest input frequency that can be tolerated without causing the output pulse to be shorter than one-half cycle of the input? (b) If spiking must be prevented but narrowing of the output pulse is permitted, which specification, delay of high-to-low transitions, t_{HL}, or delay of low-to-high transitions, t_{LH}, must be improved in order for higher input frequencies to be allowed?

Solution (a) If the desired output pulse width is T, it will be foreshortened and a spike will be created if $\frac{1}{2}f_{IN}$ is delayed by T, if $\frac{1}{4}f_{IN}$ is delayed by $3T$, or if $\frac{1}{8}f_{IN}$ is

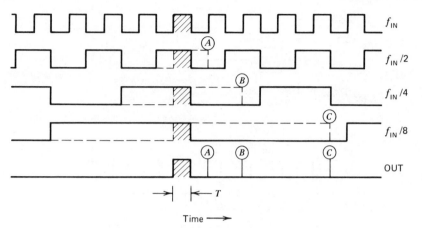

Figure 6.11 Waveforms for Example 6.1.

delayed by $7T$. These delays are shown by dashed lines at A, B, and C in Fig. 6.11. The spikes caused at the output by each delay are correspondingly labeled. The shaded area must exist in each waveform to produce a full-width output pulse. The total delays of these three waveforms are 30, 79(30 + 49), and 203(79 + 124) nsec, respectively. Therefore, the frequencies producing foreshortening and spiking due to each waveform are, respectively, $1/(2 \times 30$ nsec), $2/(2 \times 79$ nsec), and $7/(2 \times 203$ nsec), or 16.7, 19.0, and 17.2 MHz; the problem will occur first at 16.7 MHz in the $\frac{1}{2}f_{IN}$ waveform. (Note that the factor of $\frac{1}{2}$ is due to the fact that T represents one-half the input period: $f_{IN} = 1/2T$.) (b) Since spiking is caused by a delay in the one-to-zero (high-to-low) transition, t_{HL} must be reduced.

Figure 6.8 shows two different ways to connect the fourth stage. At (a), the J and K inputs are driven by a gate which senses each Q output. In order to reduce gate size, the gate in (b) senses only the Q output of the previous stage and the output of the gate in the previous stage. The difference is more apparent in longer dividers, such as in Figs. 6.12 and 6.13. With the method shown in Fig. 6.12, it is apparent that the number of gate inputs could become quite large. The first stage also must drive a very heavy load. However, when the last stage is about to change state, the first stage has just gone to (Q equal to) one, and this can be rapidly sensed by the last stage,

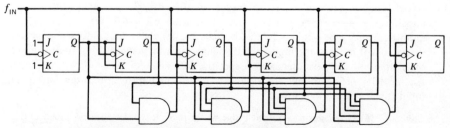

Figure 6.12 Synchronous divider with full parallel enable.

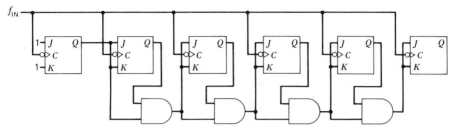

Figure 6.13 Synchronous divider, serial enable circuit.

whereas, in Fig. 6.13 the one must propogate through many gates. Therefore, a configuration such as shown in Fig. 6.13 can limit speed. Often a combination of the two techniques is used, as we shall see.

6.3 PRESETTABLE DIVIDERS

The presettable divider is used to obtain a variable, controllable divide ratio. A typical element of the divider is a J-K flip-flop with asynchronous (not clocked) set (S) and reset (R) capability, as shown in Fig. 6.14. The set function is sometimes called preset (P).

A simple presettable divider is shown in Fig. 6.15. It counts asynchronously and the final number F sensed by the AND gate is binary 1001. When F is reached, a pulse is generated which sets or resets the divider stages, depending on the preset number P. Only one function, R or S, is needed and, if a given flip-flop has only one of these, it can be redefined to equal the others by redefining (interchanging) the Q and $\overline{Q}$ outputs. If the final state of a stage is one, then it need never be set, but only reset, and this would happen only if the corresponding preset digit were zero. Thus, a stage that is one at final count has a reset input connected to an inverted preset digit. Likewise, a

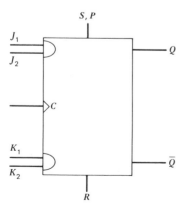

Figure 6.14 J-K flip-flop with asynchronous set and reset.

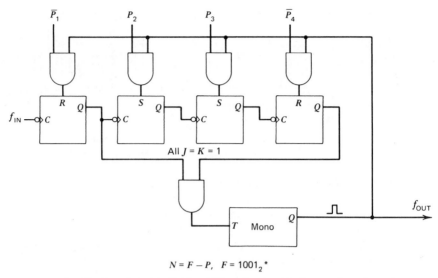

$$N = F - P, \quad F = 1001_2 *$$

Figure 6.15   Simple presettable divider. (•The subscript 2 indicates a base-2 number.)

stage that is zero at final count has a set input connected to a noninverted preset digit during presetting.

No clock pulses are required to go from the final count F to the present number P so the total number of input pulses required for each output pulse is

$$N = F - P. \tag{6.1}$$

The monostable is used because, without it, the first or fourth bistable might be preset, and thus cause the presetting pulse to end, before the others preset.

Another asynchronous divider is shown in Fig. 6.16. Here, four bistables are formed into a four-stage binary, or a hexadecimal, divider by MSI techniques.[7] A zero at the preset input (PRE = 1) causes the stages to go to the value given by the data inputs (A through D). The presetting mechanism is faster than in Fig. 6.15. Many of the serial delays which occur in generating a preset there have been eliminated. The final count is again sensed by an AND gate, but, rather than directly presetting the divider, this signal merely sets up the circuit for a preset. The next input pulse generates the preset by triggering the preset flip-flop (PFF). The next input pulse resets the PFF and is not counted due to the inhibiting action of the preset signal, PRE. Assuming that the sensed inputs to the AND gate arrive on time, the presetting mechanism limits the minimum input period to be equal to the sum of the propagation delay of PFF and the time allowed between the removal of the PRE and the next clock (enable time) or, approximately, to the minimum required PRE pulse width, whichever is greater. For the circuit of Fig. 6.15, the propogation delay through the divider plus AND gate plus

7   For example, Type 54197.

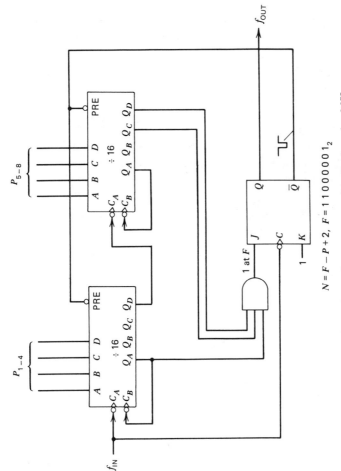

Figure 6.16 High-speed asynchronous presettable divider using MSI.

$$N = F - P + 2, \quad F = 11000001_2$$

MONO plus the MONO pulse width (the minimum width is set by the PRE requirement), plus the delay through another AND gate, plus the divider enable time must be less than the input period.

In Fig. 6.16, one input is required immediately after the final sensed count F to set PFF and another is required to reset it before counting can start from the preset value P. Thus the number of input counts for each output pulse is

$$N = F - P + 2. \tag{6.2}$$

It is necessary that the AND-gate output occur before the next input pulse in order that the PFF trigger on the desired clock transition. Whether this will occur is determined by adding the propogation delays of importance. The more consecutive ones which appear, beginning with the least significant digits in F, the less problem will be caused by propogation delay. A final number like 10000_2 was preceded by 01111_2.* The one in the most significant place cannot be sensed until the change has propogated all the way down the chain of bistables. By the time this occurs, the first digit may already be changing. This is illustrated in Table 6.1. A final count such as 10001 gives another input period for the propogation to the last digit to occur, and the number 10011 gives three more.

Table 6.1 Propagation of State Changes
in an Asynchronous Divider

┌─────── Least significant bit	
$\dfrac{01111}{01110}$	clock
01100	
01000	
$\dfrac{00000}{10001}$	clock
10001	
10001	
$\dfrac{10001}{10000}$	clock
10010	
10010	

Figure 6.17 shows a synchronous presettable divider.[8] Each MSI package contains four binary dividers. Enable-P (EN-P) and enable-T (EN-T) must both be one to permit the state to change when the clock goes positive. Enable-T must be one, and all stages within that divider must be one (decimal 15_{10} = binary 1111_2), to allow a one at the carry output. Internally, synchronization is achieved with a circuit like that of Fig. 6.12. However, the

---

* The subscript following the number indicates the base.

8 For example, Type 54161.

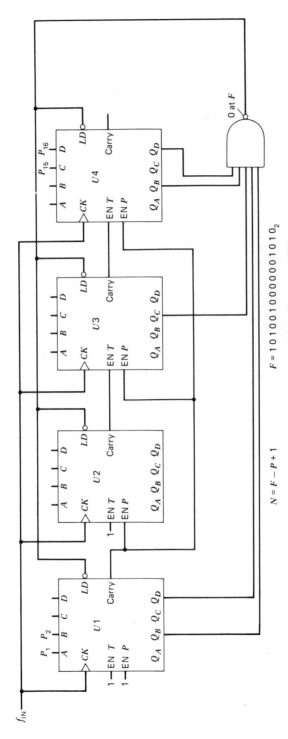

$$N = F - P + 1$$ $$F = 1010010000001010_2$$

Figure 6.17 Synchronous presettable divider using MSI.

four packages are interconnected with an arrangement more like that of Fig. 6.13. U1 counts each input cycle because enable-T and enable-P are both one. Each time state 15 is reached, the carry output becomes one. This propogates to each enable-P of the subsequent stages, permitting them to count the next clock if their enable-T is also one. U2 will thus count every time U1 goes from 15 to zero, in normal fashion for binary counting. When U2 reaches 15, a carry output will occur because enable-T is one for U2. When U1 next reaches 15, the enable-P input to U2 and U3 will become one. At this time U1 through U3 have all enable inputs at one so the next clock causes all three stages to count. The process is repeated until U3 reaches 15, then U2 reaches 15, then U1 reaches 15. At this point both enable inputs to U4 are one and all four hexidecimal dividers count on the next clock. Thus, no stage can count unless each previous stage has reached its final count, 15. This information is transmitted through the stages, in a manner similiar to that for the circuit of Fig. 6.13, except the carry output of the first stage goes to all stages in parallel. When the count is, from left to right, 15-14-15-n_4, the next clock pulse produces 0-15-15-n_4. n_4 will not change until the next clock pulse after U1 reaches 15. Then the information that U1 = 15 will be transmitted directly from U1 to U4 via enable-P. The fact that U2 had reached 15 will then have had 16 input pulse periods to arrive at U4 via the serial path through U3.

The final number can be sensed and sent to the load (LD) input of each IC. The next clock pulse then causes each IC to preset, thus removing the output at the gate and the load inputs.

After final (sensed) count is reached, one pulse is required to preset. The total number of input pulses required to produce one output is therefore

$$N = F - P + 1. \tag{6.3}$$

EXAMPLE 6.2

Problem   Fill in the missing blanks in the following table. The figures indicate the type of divider arrangement, not the exact configuration.

Refer to Figure	Divide by N	Preset No. P	Final No. F	Base
(a) 6.15	400	200		10
(b) 6.15	0111110		1100011	2
(c) 6.16		450	10,000	10
(d) 6.16	63	26		10
(e) 6.17	10,000	1,000		10
(f) 6.17	10101		11010	2

Solution

$N = F - P$	(a) $F = 600_{10}$	(b) $P = 0100101_2$
$N = F - P + 2$	(c) $N = 9552_{10}$	(d) $F = 87_{10}$
$N = F - P + 1$	(e) $P = 10,999_{10}$	(f) $P = 110_2$

EXAMPLE 6.3

Problem In Fig. 6.16, assume the ÷16's have propogation delays of 10 nsec for either transition (t_{HL} or t_{LH}) from C_A to Q_A and the other interstage delays, from C_B to Q_B, from Q_B going to zero to Q_C changing, etc., are 20 nsec for transitions to one (t_{LH}) and 25 nsec for transitions to zero (t_{HL}). The AND gate has a propogation delay of 5 nsec and the setup time (t_{SU}) from arrival of a signal at J of PFF to the arrival of the clock there must be 10 nsec. The above are maximum times. What is the highest input frequency which will certainly allow presetting to begin on the correct clock pulse for the following final counts?

(a) 1 1 1 1 1 1 1 1

(b) 1 0 0 0 0 0 0 1

(c) 1 0 0 0 0 0 0 0,

from most to least significant.

Solution Refer to Fig. 6.18. Let the clock edge that triggers the final number occur at $t = 0$. Then the time t when the system is ready for the next clock, which triggers the J-K flip-flop, is given in terms of the clock period T and frequency f as follows.

(a) Due to LSB:

$$t = t_{CA\downarrow - QA\uparrow} + t_G + t_{SU} < T,$$

$$10 \text{ nsec} + 5 \text{ nsec} + 10 \text{ nsec} < 1/f,$$

$$f < \frac{1}{25 \text{ nsec}} = 40 \text{ MHz.}$$

Due to 2LSB:

$$t = -T + t_{CA\downarrow - QA\downarrow} + t_{CB\downarrow - QB\uparrow} + t_G + t_{SU} < T,$$

$$10 \text{ nsec} + 20 \text{ nsec} + 5 \text{ nsec} + 10 \text{ nsec} < \frac{2}{f},$$

$$f < \frac{1}{22.5 \text{ nsec}} = 44.4 \text{ MHz.}$$

Due to 3LSB:

$$t = -3T + t_{CA\downarrow QA\downarrow} + t_{CB\downarrow - QB\downarrow} + t_{QB\downarrow - QC\uparrow} + t_G + t_{SU} < T,$$

$$10 \text{ nsec} + 20 \text{ nsec} + 25 \text{ nsec} + 10 \text{ nsec} + 5 \text{ nsec} < 4T = \frac{4}{f}$$

$$f < \frac{4}{70 \text{ nsec}} = 57.1 \text{ MHz.}$$

ANSWER: 40 MHz.

(b) Due to MSB:

$$t = -T + t_{CA\downarrow - QA\uparrow} + t_{CB\downarrow - QB\downarrow} + t_{QB\downarrow - QC\downarrow} + t_{QC\downarrow - QD\downarrow}$$

$$+ t_{CA\downarrow - QA\downarrow} + t_{QA\downarrow - QB\downarrow} + t_{QB\downarrow - QC\downarrow} + t_{QC\uparrow - QD\uparrow} + t_G + t_{SU}$$

$$= 2(10 \text{ nsec}) + 5(25 \text{ nsec}) + 20 \text{ nsec} + 5 \text{ nsec} + 10 \text{ nsec} - T$$

$$= 180 \text{ nsec} - T < T$$

$$f < \frac{2}{180 \text{ nsec}} = 11.1 \text{ MHz.}$$

(c) Due to MSB:

$$t = 180 \text{ nsec} < T,$$

$$f < 5.56 \text{ MHz.}$$

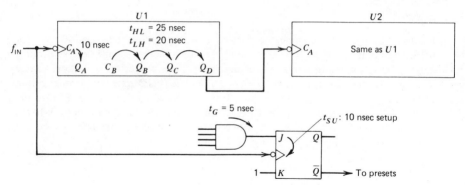

Figure 6.18   Divider for Example 6.3.

6.4   PRESCALERS AND PULSE SWALLOWERS

Prescalers divide by fixed ratios. When the divider input frequency is too high to permit proper operation of a presettable divider, a prescaler can be employed to lower that frequency. Prescalers can often operate at higher frequencies because they do not have to allow for the delays involved in presetting. A few high-speed prescaler stages can permit lowering of the speed requirements for many subsequent stages. The disadvantage is that, for a given resolution (step size), the reference frequency must be lowered when a prescaler is employed. The total divide ratio is

$$N = N_p N_v \tag{6.4}$$

when N_p is the prescaler divide ratio and N_v is the varaible (presettable) divider ratio. Since N_p is constant, the smallest change in divide number occurs when N_v is changed by 1 and is

$$\Delta N = N_p. \tag{6.5}$$

Thus, the step size is

$$\Delta f = N_p f_{\text{REF}}. \tag{6.6}$$

So, for a given resoltuion, f_{REF} must be lowered by N_p, and with the lower value of f_{REF} comes lower allowable loop bandwidth. Quite often lower bandwidth is undesirable. The solution in that case may be a pulse-swallower circuit.[9, 10]

The pulse swallower allows generation of the multiples of Δf which lie between the steps given by Eq. (6.4), that is,

$$(N_p N_v + 1)f_{\text{REF}}, (N_p N_v + 2)f_{\text{REF}}, \ldots, (N_p N_v + N_s)f_{\text{REF}}, \ldots.$$

It does this by "swallowing" N_s input pulses. The overall divider acts like a variable divider preceded by a prescaler, but N_s input pulses are somehow lost, or swallowed, making the input frequency higher by N_s each reference

9 Motorola Data Sheet, MC12012.
10 Nichols.

period. The circuit that allows the pulse swallower to swallow pulses is the variable-modulus prescaler (VMP). The VMP is capable of dividing by either N_{ps} or $N_{ps} + 1$. The reference frequency is not changed when a VMP is introduced. It replaces divider stages that would otherwise have been part of a standard presettable divider.

How can the VMP lower the speed requirement for the subsequent divider stages while, at the same time, maintaining the same resolution as an ordinary presettable divider? The reason that a presettable divider does not necessarily lower the maximum frequency to subsequent stages is that, at preset time, the allowable transition time may become much smaller than it otherwise is. Depending on the preset number, a divider stage may provide a clock transition to the next stage immediately after preset and, depending on the final count, it may have changed state just before preset.* For example, in Table 6.2, if the final count is 12 and the preset number is 7, with the simple divider of Fig. 6.15 the 2^2 bit would be high for only one clock period. The 2^3 divider would thus see a pulse no wider than the usual output of the 2^0 divider, even though there are two divider stages between them. The VMP, on the other hand, never divides by less than its lower modulus.

Table 6.2 Binary Count Sequence

2^3	2^2	2^1	2^0	Decimal
0	0	0	0	0
0	0	0	1	1
0	0	1	0	2
0	0	1	1	3
0	1	0	0	4
0	1	0	1	5
0	1	1	0	6
0	1	1	1	7
1	0	0	0	8
1	0	0	1	9
1	0	1	0	10
1	0	1	1	11
1	1	0	0	12
1	1	0	1	13
1	1	1	0	14
1	1	1	1	15

The reason that the ratio of resolution to reference frequency must increase when a prescaler is used is that there is no way to alter the count in the bits that are represented by the prescaler. The presettability of the other stages permits the count to be shortened by skipping a group of numbers at preset. In the above example, all numbers from 12 to 15 to 0 to 6 were skipped due to preset. The second modulus of the VMP, on the other hand,

---

* Actually, regardless of the value of the final count, the preset action, which is initiated at that time, must be completed in time to permit the proper response to the next clock.

allows the VMP divide ratio to be varied without decreased spacing between output transitions. When the higher modulus, $N_{ps} + 1$, is employed, an input clock pulse is said to be swallowed because the result is as if an input pulse to a prescaler, of modulus N_p, had been eliminated. For each swallowed pulse, the input frequency must be higher by one cycle per reference period. The count length thus changes by one in the least significant bit, counting the VMP as part of the divider. There is a restriction, however, caused by the fact that the VMP swallows only one pulse for each complete cycle of the VMP. It must generally be capable of swallowing $N_{ps} - 1$ pulses in order to allow an arbitrary value from 0 through $N_{ps} - 1$ for the digits it represents in the total divide ratio. To do this, it must produce $N_{ps} - 1$ output pulses, so the minimum count of the subsequent stages must be at least $N_{ps} - 1$. Thus, the VMP can consist of only about one-half of the stages required for the minimum count. More precisely, the total count N is

$$N = N_{ps}N_2 + N_s, \tag{6.7}$$

where N_2 is the number of pulses into the variable (presettable) counter each reference period and N_s is the number of swallowed pulses per reference period. As discussed above, the capacity of the VMP is limited by

$$N_{ps} \leqslant N_2 + 1. \tag{6.8}$$

Combining expressions (6.7) and (6.8), we obtain

$$N_{ps}^2 \leqslant N + N_{ps} - N_s, \tag{6.9}$$

At the maximum value of N_s, $N_{ps} - 1$, (6.9) becomes

$$N_{ps} \leqslant \sqrt{N + 1} . \tag{6.10}$$

We have assumed that the VMP can divide by $N_{ps} + 1$ continually, if necessary. If it is prevented from doing so during some cycle of the VMP (e.g., during preset) the maximum value for N_{ps}, as given by (6.10), must be reduced, generally by about one-half count [i.e., to $N_{ps} - 0.5$, where N_{ps} is taken from (6.10)].

Figure 6.19 illustrates an entire pulse swallower, consisting of the VMP, a

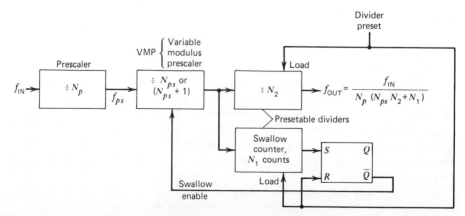

Figure 6.19 Divider with prescalar and pulse swallower.

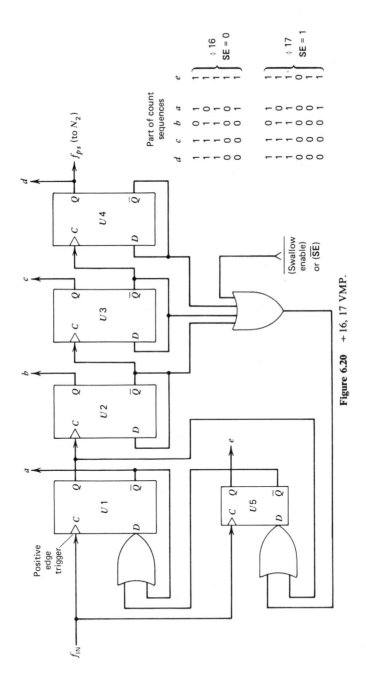

Figure 6.20 ÷16, 17 VMP.

swallow counter and an *S-R* flip-flop. The *S-R* flip-flop is reset at the same time the divider is reset and it enables the swallow, that is, it causes the VMP to divide by $N_{ps} + 1$. After N_1 counts from the VMP, the swallow counter sets the *S-R* flip-flop, preventing further swallows. Thus, N_1 represents the number of swallowed pulses.

If the divider reset pulse is very narrow, this configuration can work properly for $N_1 = 0$, since the swallow counter will, in this case, be preset to its final value and, as soon as the divider reset pulse disappears, will set the *S-R* flip-flop. Depending upon the timing within the VMP, this may prevent any pulses from being swallowed, as it should. However, it is often necessary to inhibit the swallow enable during divider preset to insure proper operation.

Figure 6.20 shows a $\div 16, 17$ VMP composed of five *D* flip-flops. Flip-flops $U1-U4$ act as an ordinary four-bit ripple counter for as long as a one exists on *Q* of flip-flop $U5$. At the count (from *d* to *a*) of 1111, the *D* input to flip-flop $U5$ goes to zero if the swallow in enabled. The next clock pulse sends the count to 0000 but also sets the $\overline{Q}$ output of $U5$ to one. This prevents $U1$ from changing states on the next clock. $U5$ does change states, however, so the subsequent clock again triggers $U1$. Thus whenever the swallow is enabled, and only when it is enabled, an input pulse is swallowed as the divider spends two periods at state 0000. This can be seen by comparing the two count sequences given in Fig. 6.20. This divider can be made very fast, but not quite as fast as the toggle rates of flip-flops $U1$ and $U5$. The other flip-flops can be made progressively slower as they are further from the input. The one-to-zero transition of the last state (at *d*) should be used to trigger the rest of the divider (N_2 in Fig. 6.19) so maximum delay, nearly one cycle at f_{ps}, can be tolerated in changing the state of the swallow enable.

The lower modulus of a VMP can be multiplied by the technique shown in Fig. 6.21. This is important if an integrated VMP is available but a higher modulus is needed. When the $\div N_e$ reaches its final count, the swallow is enabled in the VMP if the main swallow enable signal is at one. Since this occurs only every N_eth cycle of the VMP, the overall modulus is either $N_{ps}N_e$ or $N_{ps}N_e + 1$.

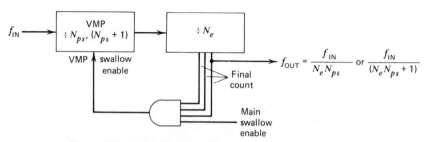

Figure 6.21   Multiplication of the (lower) modulus of a VMP.

6.5   OTHER BASES

Our discussion has related to binary dividers, but the synthesizer's divider can be composed of dividers of other bases. The most usual base, other than two, is 10. Often the divider will be composed of binary-coded-decimal (BCD) dividers because the control word is so organized. Integrated BCD dividers are readily available and a $\div 10, 11$ VMP can be obtained that toggles to 500 MHz. The organization and operation is basically as has been described for binary elements. In Figs. 6.16 and 6.17, for example, the $\div 16$ blocks are replaced by $\div 10$ blocks which recycle, and cause the next stage to count, after every 10 inputs, rather than after every 16.

6.6 PRESETTING AND OFFSETTING

Often the synthesizer control word* has a specified relationship to synthesizer frequency or to some frequency related to the synthesizer frequency (e.g., the received frequency for a receiver where the synthesizer acts as a LO and is thus offset by one or more IF frequencies). It then becomes advantageous to choose an architecture and a final count which will permit easy translation from the control number to the required synthesizer frequency. In this section, we study the mathematical relationship between the various divide numbers N, the final count F, the preset number P, and the control word C. Figure 6.19 will serve as a general model of the divider.

$$N = N_p(N_{ps}N_2 + N_1).^\dagger \qquad (6.11)$$

The value of N_2 is

$$N_2 = F_2 - P_2 + R_2, \qquad (6.12)$$

where R_2 is the number of input (to $\div N_2$) pulse periods from the final count to preset. R_2 equals zero for Fig. 6.15, one for Fig. 6.17, and two for Fig. 6.16.

Assuming that each VMP output pulse advances the swallow counter by one and that pulse swallowing is stopped when the count reaches F_1, the value of N_1 is given by

$$N_1 = F_1 - P_1. \qquad (6.13)$$

Combining Eqs. (6.11)–(6.13), the divide ratio is

$$N = N_p N_v, \qquad (6.14)$$

where

$$N_v = F - P + N_{ps}R_2, \qquad (6.15)$$

$$F = F_2 N_{ps} + F_1, \qquad (6.16)$$

and

$$P = P_2 N_{ps} + P_1. \qquad (6.17)$$

F and P are simply the final and preset numbers, respectively; when the base is the same for the pulse swallower as for the variable divider, they apply to the swallow counter as if it were the least significant stages of the overall counter.

---

* In other words, the number represented by the binary signals that select the synthesizer's frequency.
† With no prescaler, $N_p \to 1$. With no pulse swallower, $N_{ps} \to 1$ and $N_1 \to 0$.

EXAMPLE 6.4

Problem In Figure 6.19, the prescalar modulus N_p is 2. N_{ps} is decimal 10. The swallow counter is preset to 2 and it sets the *S-R* flip-flop at a count of 4. N_2 is a divider such as shown in Fig. 6.17. It is preset to 10_{10} (decimal) and the final number is 2002_{10}. If f_{OUT} is 100 Hz, what is f_{IN}?

Solution Refer to Fig. 6.19.

$N_1 = 4 - 2 = 2$

$N_2 = F - P + 1 = 2002 - 10 + 1 = 1993$

$N_p = 2$

$N = N_p(N_{ps}N_2 + N_1) = 2\,(10(1993) + 2) = 39{,}864$

$f_{in} = f_{out}\,(39{,}864) = 3.9864$ MHz

If the frequency is to change in the opposite direction from the control number C, the divide number can be described as

$$N_v = O - C, \tag{6.18}$$

where O is an offset. This can be combined with Eq. (6.15) to give

$$O - C = F - P + N_{ps}R_2. \tag{6.19}$$

The control lines can then be used directly for preset, giving

$$P = C, \tag{6.20}$$

so the final count required for a divide number given by Eq. (6.18) is

$$F = O - N_{ps}R_2. \tag{6.21}$$

In the somewhat more common case, where the frequency must advance with the control number, we have

$$N_v = O + C, \tag{6.22}$$

$$O + C = F - P + N_{ps}R_2. \tag{6.23}$$

The inverse of each bit can easily be used for preset. The preset number is then

$$P = N_m - C, \tag{6.24}$$

where N_m is the largest possible number that has the same quantity of digits as C. For binary control, N_m is all ones, so $N_m - C$ is the one's complement, easily obtained by inverting all bits. For BCD, N_m is all nine's; $N_m - C$ is the nine's complement and is obtained by subtracting each BCD digit from nine. The advantage of using these complements is that they are created digit by digit, with no carries. Combining Eqs. (6.23) and (6.24), we obtain the final count corresponding to Eq. (6.22) as

$$F = O + N_m - N_{ps}R_2. \tag{6.25}$$

The offset will often cause an extra stage to be required. In choosing O or N_{ps} or R_2, one must bear in mind the importance of the distributions of one's in F as discussed in Section 6.3.

EXAMPLE 6.5

Problem In a configuration similar to Fig. 6.16, the binary control word is a receiver frequency in MHz. The synthesizer provides the LO which must be 60 MHz higher. (a) What is the minimum processing required of the control word? (b) If the maximum receiver frequency is 120 MHz, how many binary stages are required, and (c) what is the final count?

Solution

(a) Obtaining the one's complement is the minimum processing.

(b) (120 MHz + 60 MHz)/1 MHz = 180. This is the divide ratio required to divide the highest synthesizer output frequency down to the highest allowed reference frequency which equals the smallest increment in the control word. Eight stages are required, since eight stages have a divider capacity of 256 while seven stages have a capacity of only 2^7 or 128.

(c) In Eq. (6.25), $O = 60$, $N_m = 127$, $N_{ps} = 1$, $R_2 = 2$. $F = 60 + 127 - 2 = 185$. The minimum preset number is $127 - 120 = 7_{10} = 111_2$. The count proceeds from there to the final count, $185_{10} = 10111001_2$. Suppose that, in this problem, the highest receiver frequency were 67 MHz. Then the minimum preset would be $127 - 67 = 60_{10} = 111100_2$. The final number is still 10111001. The maximum divide number is $67 + 60 = 127$, which could be handled by seven stages (e.g., by starting at zero and going to 125, two counts being used in presetting). Thus an extra stage is required in order to simplify the presetting in this case, but it is probably worth it.

EXAMPLE 6.6

Problem Do Example 6.5 above, but assume the divider of Fig. 6.16 is preceded by a pulse swallower as shown in Fig. 6.22.

Solution

(a) Generation of a one's complement is the minimum processing.

(b) Eight stages are required, six in N_2 and two in VMP during normal operation, when the lower modulus is being employed.

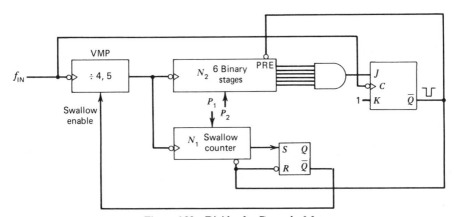

Figure 6.22 Divider for Example 6.6.

(c) $F = 60 + 127 - 2(4) = 179$. Here Eq. (6.25) was used.
$$179_{10} = 10110011_2.$$

This is divided into final numbers for N_1 and N_2, F_1 and F_2, respectively:
$$F_2 = 101100, F_1 = 11.$$

6.7   OTHER LOGIC CIRCUITS USED IN SYNTHESIZERS.

A second method for increasing the time available for presetting the stages of
an asychronous divider is the use of what may be called an auxiliary counter.
This begins to count when the final number is reached in the main divider.
While it is counting, the main divider is inhibited. The decoded count of the
auxiliary counter may then be used to control the preset sequence in the main
divider. When the auxiliary counter returns to its intial value, the main
divider begins to divide again.

Memory circuits are sometimes included with the divider logic so the
command word need not be present continually. The loading of the memory
may be synchronized with presetting of the divider in order to prevent a
transient that would result if part of the previous control word were used,
along with part of the new control word, for presetting.

If the frequency of a signal used for heterodyning in the feedback path
(f_{mix} in Fig. 2.8c). is switched, it too may be synchronized.[11]

6.8   PRECAUTIONS

The logic design for a synthesizer must be carefully checked to insure that all
of the many logic elements are operating within their ratings at all times. The
main specifications that require detailed checking are fan-out and timing
restrictions.

If all elements used are of the same logic family, checking for fan-out can
be relatively easy. The designer should be aware, however, that all inputs to
elements of the same family do not always represent one standard load. Since
the speed requirement will not be the same for every part of the logic, the use
of more than one logic family is quite feasible. This may be advantageous,
compared to using all elements from one family that is fast enough for the
most demanding part of the circuit. High-speed circuits dissipate more power
and, at the highest speeds, require special layout methods,[12] and all MSI logic
functions are not available in all families. In general, for each output of a
logic element, the total load, for both the zero and one output states, must be
computed and compared to the drive capability of the driving logic element.

Proper timing is even more difficult to check, but the range of arrival of
each important transition at each logic element input must be computed to

11   Egan, "LO's Share Circuitry . . . ," pp. 60 and 65.
12   Blood.

insure that it occurs within an acceptable window. For example, a flip-flop data input (J, K, or D) may have a specified setup time, indicating how long before the clock transition the data must arrive, and a specified hold time, indicating how long after the transition the data must remain. Its clock will have a specified minimum pulse width or maximum frequency. Other inputs, like set and reset, will also have timing specifications.

Other potential problem areas are heat and noise. Logic can generate considerably heat and its removal must be planned, particularly for high-speed logic and in a confined region, or failures may result. The switching transients of logic circuits generate considerably noise which can be coupled to other circuits through the fields created, through ground currents, or on power supply lines.[13-16] Within the synthesizer, this noise can enter the analog circuitry and cause FM on the synthesizer output. External to the synthesizer, it can interfere with other electronics. Care should be taken in planning the grounding and shielding of the logic and the filtering of lines which must enter this high-noise region.

6.9   INTERFACING WITH THE ANALOG (RF) INPUT

Since the variable divider will usually be driven by a sinusoidal signal, a transition to logic voltage levels is required. A circuit is desired which needs little input swing and is nevertheless tolerant of a wide range of input levels. One useful circuit for converting to TTL levels is shown in Fig. 6.23a. We analyze it as follows.

D is a Schottky diode which prevents Q from saturating and thus improves switching speed. When V_p is small, the voltage at the base of Q will be approximately 0.7 V (assuming a silicon transitor), so the current through R_y will be

$$I_y \approx \frac{4.3 \text{ V}}{R_y} .$$

(6.26)

Assume that the time constant $R_x C$ is long compared to the period T, so the voltage across C is constant, and that V_p is large compared to the minimum swing required at the base to switch the transistor (a few tenths of a volt). When Q is barely cutting off, the peak current in each direction through R_x will be

$$I_p \approx \frac{V_p}{R_x} .$$

(6.27)

The minimum peak current to drive the transistor into cutoff and into saturation is

$$(I_p)_{\min} \approx \frac{5 \text{ V}}{2\beta R_L}$$

(6.28)

---

13 Morrison.   **15** Manassewitsch, pp. 154–225.
14 White.   **16** Ott.

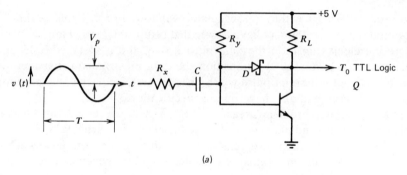

(a)

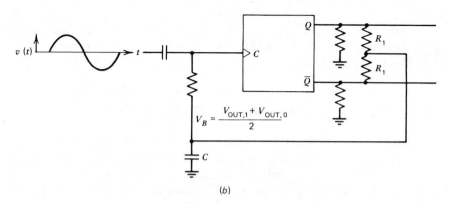

(b)

Figure 6.23 Circuits for conversion from a sinusoid to logic levels.

and it occurs if $I_p = I_y$. Then the peak input current is just sufficient to cut off the transistor by absorbing I_y and just sufficient, when added to I_y, to saturate the transistor.

As V_p grows larger, the average voltage at the base of the transistor drops. This increases the average current through R_y, which equals the average current into D and the base of Q, and also decreases the portion of the cycle during which Q is conducting. Thus higher drive levels lead to a more rectangular pulse but a shorter turn-on time for Q.

We will now compute the conditions for equal on and off times for Q when the required base input current is much smaller than its drive. Whatever charge passes through C into the transistor, during the positive part of the input swing, must be equaled by the charge that fills the capacitor when the transistor cuts off. Therefore, 50% duty cycle requires that the average input current during the positive half of the input cycle,

$$\bar{I}_x \approx \frac{(2/\pi)V_p}{R_x},$$

(6.29)

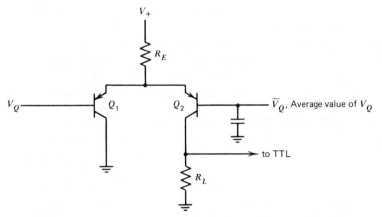

Figure 6.24 ECL to TTL translation with an emitter-coupled pair.

equal the average current entering the capacitor during the negative half of the cycle,

$$\bar{I}_y \approx \frac{4.3 \text{ V} + (2/\pi) V_p}{R_x + R_y} . \tag{6.30}$$

This will be true if the peak voltage is related to the resistance values by

$$\frac{6.75 \text{ V}}{R_y} = \frac{V_p}{R_x} . \tag{6.31}$$

The circuit in Fig. 6.23b can be used to center the input swing to an ECL flip-flop between the zero and one logic levels. Here, V_B is the average of the two opposite output states, and thus near the center of the logic swing.

An emitter-coupled pair (Fig. 6.24) is useful for translating from ECL to TTL levels when only positive supply voltages are involved. When V_Q goes low, $Q2$ cuts off; R_L must then be small enough to sink the TTL-zero-level current without excessive voltage rise. When V_Q goes high, $Q1$ cuts off and the high output level is determined by the current through R_E. This circuit can also be AC coupled at the input, like the common-emitter circuit (Fig. 6.23a), to convert a sine wave to TTL levels. Conversely, the common-emitter circuit can be used as a level translator. The emitter-coupled pair is better for high frequencies.

PROBLEMS

6.1 In Fig. 6.6, the maximum first-stage delay is 20 nsec, the maximum second-stage delay is 50 nsec, and the maximum third-stage delay is 132 nsec.

    (a) Delays are equal for positive and negative transitions and the input is a square wave. What is the highest input frequency that can be tolerated without the possibility of spikes in the output of the NAND gate in Fig. 6.7?

(b) If spiking is allowed but narrowing of the output pulse is not permitted, which specification, delay of high-to-low (one-to-zero) transitions, t_{HL}, or delay of low-to-high transitions, t_{LH}, must be improved in order for higher input frequencies to be allowed?

6.2 Draw a fifth stage for the divider of Fig. 6.8 (a) using the method shown in Fig. 6.8a and (b) using the method shown in Fig. 6.8b. Indicate connections to earlier stages by symbols used in Fig. 6.8, for example, J_4 and Q_4'.

6.3 Beginning with the most significant bit, indicate whether a set or reset input is required and whether P_i or $\overline{P}_i$ is connected for a divider like that in Fig. 6.15 if the final count is $F = 1101_2$.

6.4 In Fig. 6.16, PRE must go to zero (input to one) 25 nsec before the negative going edge of the clock, either C_A or C_B, for the divider to count that clock. This is called the enable time. PFF has a propagation delay, t_{LH}, of 12 nsec from clock to $\overline{Q}$ going high (one). Based only on these specifications, what is the maximum input frequency?

6.5 Fill in the missing blanks in the following table. The figures indicate the type of divider arrangement, not the exact configuration.

	Refer to figure	Divide by N	Preset No. P	Final No. F	Base
(a)	6.15	400		800	10
(b)	6.15	0111110	0100000		2
(c)	6.16		220	9000	10
(d)	6.16	63	26		10
(e)	6.17	10101	00111		2
(f)	6.17	10,000		10,105	10

6.6 In Fig. 6.16, assume the ÷ 16's have propagation delays of 20 nsec for either transition (t_{HL} or t_{LH}) from C_A to Q_A and the other interstage delays, from C_B to Q_B, from Q_B going to zero to Q_C changing, etc., are 30 nsec for transitions to one (t_{LH}) and 25 nsec for transitions to zero (t_{HL}). The AND gate has a propagation delay of 15 nsec and the setup time from arrival of a signal at J of PFF to the arrival of the clock there must be 10 nsec. The above are maximum times. What is the highest input frequency which will certainly allow presetting to begin on the correct clock pulse for the following final counts?

(a) 1 1 1 1 1 1 1 1,

(b)   1 0 0 0 0 0 0 1,

(c)   1 0 0 0 0 0 0 0,

from most to least significant.

6.7   What is the maximum number of binary stages in a VMP that is part of a divider with a divide range from 1000 to 2000?

6.8   In Fig. 6.19, the prescalar modulus N_p is 4. N_{ps} is 8. The swallow counter is preset to 2 and it sets the S-R flip-flop at a count of 7. N_2 is a divider such as shown in Fig. 6.17. It is preset to 100_{10} and the final number is 1004_{10}. If f_{OUT} is 10 Hz, what is f_{IN}?

6.9   In a configuration similar to Fig. 6.16, the binary control word is a receiver frequency in increments of 100 KHz. The synthesizer provides the LO which must be 21.4 MHz higher. (a) What is the minimum processing required of the control word? (b) If the maximum receiver frequency is 100 MHz, how many binary stages are required, and (c) what is the final count?

6.10   Do Problem 6.9 but assume the integrated dividers of Fig. 6.16 are BCD as is the control word.

6.11   Do Problem 6.9 but assume the divider of Fig. 6.16 is preceded by a two-stage pulse swallower ($\div 4, 5$). Part (b) will refer to the total stages including the two stages in the VMP.

7
More Complex Configurations

We have discussed the fundamental processes used in synthesis and have studied the basic components of a phase-locked synthesizer. We have also considered how these components can be combined in an elementary loop to make a synthesizer. We have looked at the linear response of the simplest type of loop and have given some consideration to the effects of a low-pass filter in the loop. Our study of the linear performance of phase-locked synthesizers would not be complete, however, without the introduction of some additional complexities, some of which have already been alluded to, which are fairly common in practice and which can offer significant design advantages.

We have noted that frequency mixing may be necessary or desirable within the loop. However, spurious signals can be introduced with this process and they can affect performance. In addition, filtering would normally follow the mixer and we must know how to introduce its effects into the block diagram.

An integrator in the loop filter may be needed to increase hold-in range and it may be used with a low-pass filter to attenuate sampling sidebands, so an understanding of the effects of various filters, including effects on the loop transient response, is necessary.

In this chapter, we will consider these added complexities and give a unified analysis of the effects of the various filter types. In addition, we will describe a different type of phase-locked synthesizer which can give an improved trade-off between small frequency increments and wide loop bandwidth.

7.1 HETERODYNING WITHIN THE SYNTHESIZER

We showed earlier (Section 2.3) that mixing may take place within the synthesizer feedback path. Figure 2.8c shows a single down conversion and Fig. 7.1 illustrates a double conversion. These are useful configurations for

168

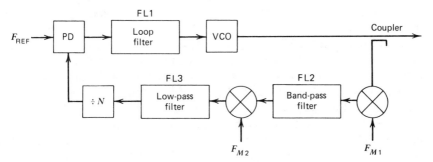

Figure 7.1 Double conversion within synthesizer.

summing several synthesized loops and, especially in microwave synthesizers, for reducing the output frequency to one which can be handled by the logic. The mixing frequencies (fixed LO's) can be generated by selecting outputs from a frequency multiplier (spectrum generator). This may be done by switching discrete filters or tuning a variable filter.[1, 2, 3A]

While it is true that several types of frequency dividers can operate as prescalers up to high microwave frequencies,[3-8] the disadvantages of prescaling noted in Section 6.4, plus the possible complexity of using these dividers and of providing RF gain to compensate for losses inherent in many of them, makes heterodyning more desirable than prescaling in many applications. As has been noted (Section 2.3), heterodyning does not change the basic mathematical representation of the loop. However, there are several things which can occur in the process and whose effects must be represented.

7.1.1 Spurious Coupling to the Output

The internal mixing frequencies, F_{M1} and F_{M2} in Fig. 7.1, are relatively strong signals that can be difficult to keep from the synthesizer output. These signals, and mixing products which they help to create, can cause serious problems in the system employing the synthesizer if they are not confined. This is also true, of course, of logic signals. The magnitude of the problem depends both upon the strength of the signal and its frequency. Special purpose synthesizers should be designed without employing strong signals at frequencies to which the using equipment is particularly sensitive, where feasible. These often include using equipment IF's and frequencies which, when mixing with the synthesizer output, produce IF's (see Section 4.3.3). In Fig. 7.1, for example, the signal at F_{M1} will be attenuated by the mixer isolation and, ideally, by the coupler directivity plus coupling factor. However, if the VCO does not present

1	Napier.	**5**	Kroupa, pp. 62–63.
2	Egan, "LOs Share . . . "	**6**	Miller.
3A	Manassewitsch, pp. 474–484.	**7**	Goldwasser.
3	Manassewitsch, pp. 352–355 and 361–370.	**8**	Bearse.
4	Penfield.		

a good output impedance match, coupled signals at frequency F_{M1} will be reflected into the output by the VCO. Additional isolation can be obtained by use of an isolator in the VCO output to improve the impedance match, a good practice anyway in order to reduce pulling, and by use of a combination of amplifiers, attenuators, and filters between the source of the offending signal and the synthesizer output. Each decibel of such amplification can provide more than 2 dB of isolation because it permits another decibel of attenuation in series and will be accompanied by more than 1 dB of reverse isolation in the amplifier. Of course, isolation in the signal path will not be effective if it is shunted by unintended signal paths created by insufficient shielding, ineffective grounding and poor power-supply isolation.

7.1.2 Spurious Coupling to the Loop

The effect of small spurious signals or modulation, present on mixing frequencies, upon the divider output can be analyzed by the methods described in Chapter 4. The synthesizer's response to the resulting modulation at the phase-detector input will then be the same (except for sign) as its response to f_{REF}. Spurious sidebands on the signals at F_{M1} and F_{M2} may be converted to AM and FM sidebands. The following discussion will apply to AM or FM on the mixing signal, or a single-sideband spur resulting in equivalent AM and FM. These will appear in the IF, regardless of whether F_{Mi} is the LO (strong) or signal (weak) input to the mixer. AM sidebands will probably be reduced by the mixer if F_{Mi} is the LO frequency but, regardless, they will be greatly attenuated in passing through the $\div N$. Here they may, however, be converted to phase modulation. This conversion can be important if the modulation originates as AM but, if the AM comes from decomposition of a single sideband, the FM component obtained in the decomposition should usually predominate.

FM will not be attenuated up to the $\div N$ input. Here it will be divided by N and new frequencies may be formed due to the previously described sampling effect. Signals leaking through the mixer, and various mixing products of the desired and undesired mixer inputs, can also be converted to FM at the divider input. Fortunately, many or all of these signals are often at multiples of the reference frequency and, therefore, fall in nulls of the hold transfer function (Fig. 5.2b) or produce only a dc offset. If modulation on the mixing frequency, at a modulation frequency of f_m, is translated to a lower modulation frequency f_{mT} by the sampling process, the loop can cause the VCO to have FM at f_{mT} such that, at the phase detector input, the two modulations tend to cancel. Thus if f_{mT} is well within the loop bandwidth, the phase deviation of the synthesizer output FM will be equal to the phase deviation at the original f_m on the mixing frequency, but at a differenct modulation frequency f_{mT}.

EXAMPLE 7.1

Problem A synthesizer has a 20-kHz loop bandwidth and a 200-kHz sample frequency. The input to the frequency divider is FM modulated at 390 kHz with a peak deviation of 1 kHz. What is the predominant sideband produced on the output? Give its amplitude and offset from the desired output.

Solution The *phase* at the output of the divider is sampled. This is significant because a different answer is obtained depending upon whether the phase or frequency at the divider output is sampled. The 390-kHz-modulation-rate phase deviation is sampled at 200 kHz, producing a component within the loop bandwidth at

$$f = 2(200) \text{ kHz} - 390 \text{ kHz} = 10 \text{ kHz}.$$

The 10-kHz component at the divider output has the same phase deviation as the 390-kHz component would have after division by the divider ratio. As a result, a 10-kHz component is produced at the synthesizer output, which, after frequency division also has this phase deviation. Because of the loop gain at 10 kHz, these two components at the divider output are forced to be almost equal and opposite in phase. Since the phase deviation due to the 390-kHz interference and the phase deviation at the synthesizer output both have the same magnitude at the divider output, they must also have the same magnitudes at the divider input. Therefore, the principal output spurs are at 10 kHz about the carier and have a magnitude of

$$20 \log_{10} \frac{1}{2} \left(\frac{1 \text{ kHz}}{390 \text{ kHz}} \right) = -57.8 \text{ dB}.$$

If the *frequency* at the divider were sampled, the modulation index at the divider output would be $(\Delta f / N)/10$ kHz. Since it is the *phase* that is sampled, the modulation index is $(\Delta f / N)/390$ kHz. In other words, it is important whether the conversion to a 10-kHz modulation frequency occurs before or after the modulation index is computed.

7.1.3   Miscounts Due to Spurious Signals

The discussion above assumes that the divider counts time-modulated zero crossings. But, what if the interfering signal distorts the divider input sufficiently to cause an additional zero crossing, as shown in Fig. 7.2. This can be prevented if the maximum slope of the interfering signal is always less than that of the desired signal because, under those circumstances, it is not possible

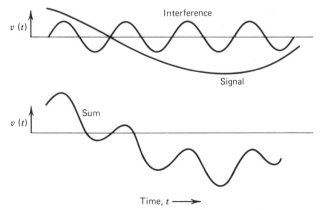

Figure 7.2   Extra zero crossing caused by interfering signal.

for the interfering signal to cause the sum signal to flatten at a zero crossing. We determine what this implies about the relative magnitudes of the desired and interfering signals as follows.

Let the desired signal be $A_D \cos \omega_D t$ and the interfering signal be $A_I \cos(\omega_I t + \theta_I)$. A zero crossing occurs where the sum of the two is zero:

$$v_s = A_D \cos \omega_D t + A_I \cos(\omega_I t + \theta_I) = 0, \tag{7.1}$$

$$\frac{A_D}{A_I} = - \frac{\cos(\omega_I t + \theta_I)}{\cos \omega_D t}. \tag{7.2}$$

If the slope is zero, we have

$$-\frac{dv_s}{dt} = A_D \omega_D \sin \omega_D t + A_I \omega_I \sin(\omega_I t + \theta_I) = 0, \tag{7.3}$$

$$\frac{A_D}{A_I} = - \frac{\omega_I}{\omega_D} \frac{\sin(\omega_I t + \theta_I)}{\sin \omega_D t}. \tag{7.4}$$

Squaring Eq. (7.2) and employing the identity $\sin^2\varphi + \cos^2\varphi = 1$, we obtain

$$\left(\frac{A_D}{A_I} \right)^2 = \frac{1 - \sin^2(\omega_I t + \theta_I)}{1 - \sin^2 \omega_D t} \tag{7.5}$$

This may be combined with Eq. (7.4) to give the conditions for zero net slope at zero voltage:

$$\left(\frac{A_D}{A_I} \right)^2 = 1 + \left[\left(\frac{\omega_I}{\omega_D} \right)^2 - 1 \right] \sin^2(\omega_I t + \theta_I). \tag{7.6}$$

In order for this condition to be impossible, $(A_D/A_I)^2$ must be larger than the maximum value for the right-hand side of Eq. (7.6), which occurs when the sine equals one. Thus, we have

$$\frac{A_D}{A_I} > \frac{\omega_I}{\omega_D} \tag{7.7}$$

as the sufficient condition for no extra counts, assuming the count occurs at the zero crossing. If the divider input is biased such that the count occurs at other than a zero crossing, the slope of the desired signal will be less and a smaller interfering signal should be capable of causing an extra count.

To insure that the interfering signal does not cause a skipped count, the peak value of the desired signal must be larger than that of the undesired signal by at least the largest peak excursion necessary to insure counting in the abscence of interference. This requirement, plus Eq. (7.7), are theoretical limits, but it is recommended that the ratio of desired to interfering signal be kept significantly greater than these would require.

EXAMPLE 7.2

Problem An interfering signal at 25 MHz enters a divider input along with a 10-MHz signal to be divided.

(a) What is the theoretical absolute maximum level for the interfering signal so it will not cause a miscount if the desired signal has an amplitude of 1-V peak and a 0.2-V peak signal is needed in the absense of interference?

(b) If the interfering-signal frequency is at 100 kHz, what is its maximum allowed level?

Solution The interfering signal must be less than 0.8 V peak. Otherwise, the desired 1-V peak signal will not reach the required ± 0.2 V to trigger the divider on some cycle. In addition, there is a frequency-dependent requirement:

(a) $A_I < (1 \text{ V peak}) \dfrac{10 \text{ MHz}}{25 \text{ MHz}} = 0.4$ V peak. This requirement predominates here.

(b) $A_I < (1 \text{ V peak}) \dfrac{10 \text{ MHz}}{100 \text{ kHz}} = 100$ V peak. But, here, the requirement that A_I be less than 0.8 V predominates.

7.1.4 The Effect of IF Filters[9]

When heterodyning is employed, the frequency conversion occurs between the VCO and the phase detector, where the loop variable is frequency deviation. We must know how this deviation is affected by IF filters (Fig. 7.1) so the effect can be introduced into the mathematical model.

Suppose the carrier frequency is in the center of the IF filter passband and the passband is symmetrical. Further, suppose the modulation index is small. The situation is pictured in Fig. 7.3a. Then the amplitude of each sideband will be changed relative to the carrier by $G_1(f_m)$, the relative gain of the filter at a separtion f_m from center. The sidebands will also receive a phase shift equal to the filter phase shift at the sideband frequency, $\theta(f_m)$ for the upper sideband and $-\theta(f_m)$ for the lower sideband. These changes in the sidebands represent a multiplication by $G_1(f_m)$ of the amplitude, and a shift of $\theta(f_m)$ in the phase, of the frequency deviation. Thus the IF filter is represented in the control system diagram by a low-pass filter with gain $G_1(f_m)$ and phase shift $\theta(f_m)$, as illustrated in Fig. 7.3b. Fig. 7.3c shows the equivalent block-diagram representations along with their input and output variables. The higher-order FM sidebands are not necessarily represented correctly by this process, so they should be small for it to be valid.

If the IF filter passband is not symmetrical with respect to the carrier frequency, there will be conversion between AM and FM that cannot always be ignored. The conversion occurs because the AM or FM sidebands are made unequal in amplitude, so they then represent a combination of AM and FM. The effect on narrow-band FM sidebands can still be represented by a low-pass filter, as in Figs. 7.3b and 7.3c, but the equivalent low-pass filter loss will be

$$G_{\text{LOW-PASS}} = \tfrac{1}{2}\big\{ G_1^2(f_m) + G_1^2(-f_m)$$

$$+ 2G_1(f_m)G_1(-f_m)\cos\big[\theta(f_m) + \theta(-f_m)\big]\big\}^{\frac{1}{2}}. \quad (7.8)$$

9 Gardner, pp. 80–84.

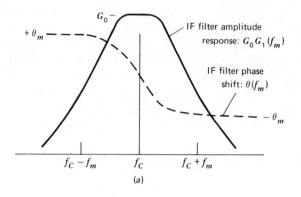

(a)

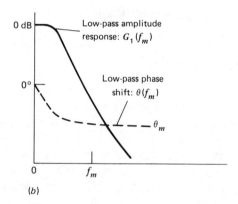

(b)

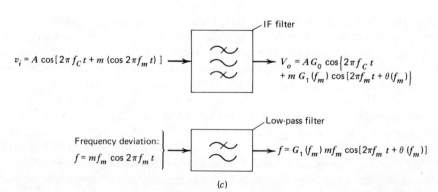

(c)

Figure 7.3 Conversion from IF filter to low-pass filter: (a) the IF filter characteristic, (b) the low-pass filter characteristic and (c) representations of the two filters.

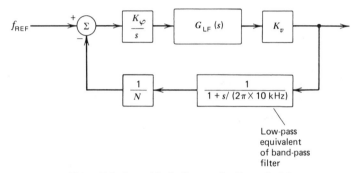

Figure 7.4 Loop block diagram for Example 7.3.

The phase shift of the equivalent low-pass will be

$$\theta_{\text{LOW-PASS}} = \tan^{-1} \frac{G_1(f_m) \sin \theta(f_m) - G_1(-f_m) \sin \theta(-f_m)}{G_1(f_m) \cos \theta(f_m) - G_1(-f_m) \cos \theta(-f_m)}. \quad (7.9)$$

EXAMPLE 7.3

Problem Draw an equivalent control loop (mathematical model) diagram for Fig. 7.1 if the loop bandwidth is 30 kHz, the low-pass filter (FL3) is flat to 100 kHz, and the band-pass filter is a single-pole LC filter, centered at the frequency passing through it, with a 20-kHz 3-dB bandwidth. The band-pass center frequency is much greater than 20 kHz (so the filter is symmetrical).

Solution The diagram is shown in Fig. 7.4. The band-pass filter has a significant effect on loop performance since the equivalent low-pass filter has its corner well within the loop bandwidth.

7.1.5 Dynamic Range With IF

Due to the number of components in series between the VCO output and the divider input, significant variations of divider RF input power can occur with temperature and frequency and from unit to unit. Therefore, a wide dynamic range at the input to the divider or into a preceding amplifier becomes important. Limiters or limiting amplifers may thus be necessary near the divider input.[10]

7.2 HIGHER-ORDER LOOPS

So far, we have discussed (Section 3.2) the responses of a simple loop, whose Bode gain plot is sketched in Fig. 7.5*a*, and the response of the same loop with a lag filter (Fig. 7.5*b*). The lag filter is an essential part of most phase-locked synthesizer loops because the reference frequency is usually not very high

---

10 "A 1-GHz prescaler . . . "

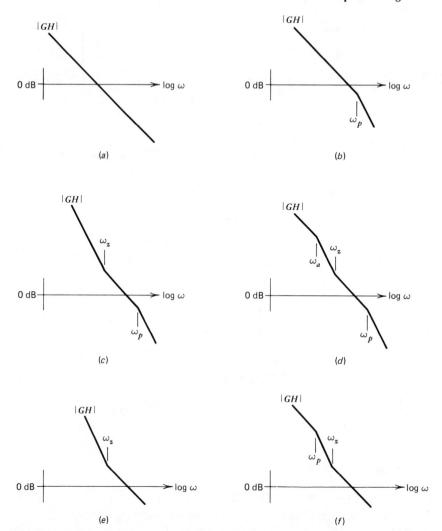

Figure 7.5 Open-loop gain for various phase-locked loops (tangential approximations): (*a*) first-order, type-1 (no filter); (*b*) second-order, type-1 (lag filter); (*c*) third-order, type-2 (integrator and lead lag); (*d*) third-order, type-1 (imperfect integrator and lead-lag); (*e*) second-order, type-2 (integrator plus lead filter); (*f*) second-order, type-1 (lag lead filter).

compared to the loop bandwidth, so the filter must be low enough in frequency that it will have an appreciable effect on the loop response. The most common additional singularities are a low-frequency zero and a very-low-frequency pole (Fig.7.5*d*), essentially at zero frequency in many cases (see Fig. 7.5*c*) so that the hold-in range and acquisition range become theoretically infinite. That is, for the simplified model, they are infinite, although circuit limitations such as, for example, finite VCO tuning range, will cause them to be finite. Unfortunately, the low-frequency pole creates a third-order loop, so we cannot rely on the easily available, well organized results for second-order loops. To understand the performance of this third-order loop, however, we can study simpler loops which have similar open-loop gain and phase profiles near ω_L, the frequency at unity open-loop gain. Such similarities can exist in the loops we have already considered and the second-order, type-2 loop (Fig. 7.5*e*). It, and the second-order, type-1 loop with lag-lead filter (Fig. 7.5*f*) are widely discussed in the phase-locked loop literature because of their applicability to phase-locked receivers, where they permit wide acquisition and hold-in ranges simultaneously with narrow bandwidth.[11,12] However, they are practically useless for many frequency synthesizers because they do not low-pass the reference-frequency components that inevitably appear in the phase-detector output. Nevertheless, the well-rounded designer of phase-locked loops should be familiar with them. They can help him to understand the performance of the third-order, type-1 loop (Fig. 7.5*d*), which is often used in practice, rather than the loop of Fig. 7.5*c*, and which is similar to the lag-lead loop if the last pole of the third-order loop is high enough in frequency.

The loop filters which correspond to each of the loop gain shapes in Fig. 7.5 are shown in Fig. 7.6.

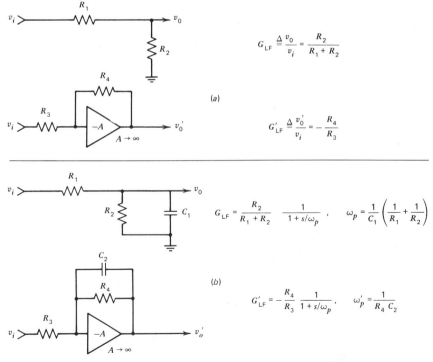

$$G_{LF} \triangleq \frac{v_0}{v_i} = \frac{R_2}{R_1 + R_2}$$

$$G'_{LF} \triangleq \frac{v'_0}{v_i} = -\frac{R_4}{R_3}$$

(a)

$$G_{LF} = \frac{R_2}{R_1 + R_2}\frac{1}{1 + s/\omega_p}, \qquad \omega_p = \frac{1}{C_1}\left(\frac{1}{R_1} + \frac{1}{R_2}\right)$$

(b)

$$G'_{LF} = -\frac{R_4}{R_3}\frac{1}{1 + s/\omega_p}, \qquad \omega'_p = \frac{1}{R_4 C_2}$$

Figure 7.6 Loop filters corresponding to Fig. 7.5: (*a*) "no filter," (*b*) lag filter, (*c*) integrator and lead lag, (*d*) imperfect integrator and lead-lag (for $\omega_a \ll \omega_z$), (*e*) integrator and lead, (*f*) lag-lead filter.

11 Gardner, pp. 8–10.
12 Viterbi, pp. 59–64.

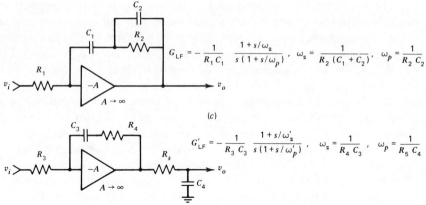

$$G_{LF} = -\frac{1}{R_1 C_1} \frac{1+s/\omega_z}{s(1+s/\omega_p)} , \quad \omega_z = \frac{1}{R_2(C_1+C_2)}, \quad \omega_p = \frac{1}{R_2 C_2}$$

$$G'_{LF} = -\frac{1}{R_3 C_3} \frac{1+s/\omega'_z}{s(1+s/\omega'_p)} , \quad \omega_z = \frac{1}{R_4 C_3}, \quad \omega_p = \frac{1}{R_5 C_4}$$

(c)

same as (c) except

$$G_{LF} \approx -A_0 \frac{(1+s/\omega_z)}{(1+s/\omega_a)(1+s/\omega_p)} , \quad \omega_a = \frac{1}{A_0 R_1 C_1},$$

refer to (c) $\quad \omega_z = \frac{1}{R_2(C_1+C_2)}, \quad \omega_p = \frac{1}{R_2 C_2}$

$$G'_{LF} \approx -A_0 \frac{(1+s/\omega'_z)}{(1+s/\omega_a)(1+s/\omega'_p)} , \quad \omega_a = \frac{1}{A_0 R_3 C_3},$$

$$\omega_z = \frac{1}{R_4 C_3}, \quad \omega_p = \frac{1}{R_5 C_4}$$

(d)

$$G_{LF} = -\frac{1}{R_1 C} \frac{1+s/\omega_z}{s}, \quad \omega_z = \frac{1}{R_2 C}$$

(e)

$$G_{LF} = \frac{1+s/\omega_z}{1+s/\omega_p}, \quad \omega_z = \frac{1}{R_2 C_1}, \quad \omega_p = \frac{1}{(R_1+R_2)C_1}$$

$$G'_{LF} = \frac{R_4+R_5}{R_3} \frac{1+s/\omega'_z}{1+s/\omega'_p}, \quad \omega'_z = \frac{1}{\frac{R_4 R_5}{R_4+R_5}C_2}, \quad \omega'_p = \frac{1}{R_5 C_2}$$

(f)

Figure 7.6 (cont.)

Table 7.1. Equations for Second Order Loop:

$D \triangleq s^2 + 2\zeta\omega_n s + \omega_n^2$; $\omega_0 \triangleq \lim_{s\to 0} sGH(s)$ (Eq. 3.10)

	General Equations	Lag-Lead Filter	Lag Filter	Integrator plus Lead
Loop filter: $\dfrac{G_{LF}}{K_{LF}} =$	$\omega_{pp}\dfrac{1+s/\omega_z}{\omega_p+s}$ (7.10)	$\dfrac{1+s/\omega_z}{1+s/\omega_p}$ (7.15)	$\dfrac{1}{1+s/\omega_p}$ (7.21)	$\dfrac{1+s/\omega_z}{s}$ (7.25)
ω_{pp}		ω_p		1
Response to reference modulation (see Fig. 3.7a): $\dfrac{f_{OUT}}{f_{REF}} =$	$K_F\dfrac{\omega_{pp}}{\omega_z}\dfrac{s+\omega_z}{D}$ (7.11)	$K_F\dfrac{\omega_p}{\omega_z}\dfrac{s+\omega_z}{D}$ (7.16) $= N\omega_n\dfrac{\left(2\zeta - \dfrac{\omega_n}{\omega_0}\right)s+\omega_n}{D}$ (7.17)	$\dfrac{K_F\omega_p}{D}$ (7.22) $= \dfrac{N\omega_n^2}{D}$ (7.23)	$\dfrac{K_F}{\omega_z}\dfrac{s+\omega_z}{D}$ (7.26) $= N\omega_n\dfrac{2\zeta s+\omega_n}{D}$ (7.27)
Response to VCO Modulation (See Fig. 3.7a): $\dfrac{f_{OUT}}{f_1} =$	$\dfrac{s(s+\omega_p)}{D}$ (7.12)	$\dfrac{s(s+\omega_p)}{D}$ (7.12) $= \dfrac{s(s+\omega_n^2/\omega_0)}{D}$ (7.18)		$\dfrac{s^2}{D}$ (7.28)
Velocity constant: $\omega_0 =$		$\dfrac{K_F}{N}$ (7.19)		∞ (7.29)
Natural frequency: $\omega_n =$	$\sqrt{\dfrac{K_F}{N}\,\omega_{pp}}$ (7.13)	$\sqrt{\omega_0\omega_p}$ (7.20)		$\sqrt{\dfrac{K_F}{N}}$ (7.30)
Damping factor: $\zeta =$	$\dfrac{1}{2}\left(\dfrac{\omega_p}{\omega_n} + \dfrac{\omega_n}{\omega_z}\right)$ (7.14)		$\dfrac{1}{2}\dfrac{\omega_p}{\omega_n}$ (7.24)	$\dfrac{1}{2}\dfrac{\omega_n}{\omega_z}$ (7.31)

7.2.1   The General Second-Order Loop[13,14]

We will now consider the loop with lag-lead filter, corresponding to Fig. 7.5f, and will show how the loops of Figs. 7.5b and 7.5e can be considered special cases of it. The equations discussed below appear in Table 7.1.

The filter characteristic corresponding to Fig. 7.5f may be written as Eq. (7.10). We write it in this form to give the flexibility of having constant DC gain (by letting ω_{pp} and ω_p be equal) or keeping the high-frequency gain constant as ω_p is lowered toward zero (by defining ω_{pp} as a constant). The closed-loop transfer functions will then be given by Eqs. (7.11) and (7.12), where the natural frequency and damping factor are defined by Eqs. (7.13) and (7.14). K_F is the forward gain divided by the normalized filter function of Eq. (7.10), at $\omega = 1$.

7.2.1.1   Lag-Lead Filter

We now set the normalized filter gain to unity at DC by letting $\omega_{pp} = \omega_p$. The filter function is then given by Eq. (7.15). Equations (7.11)–(7.13) then become Eqs. (7.16)–(7.20). The damping factor continues to be given by Eq. (7.14). A Bode plot of open-loop gain, $|GH(\omega)|$, is shown in Fig. 7.7 along curve 2, 3, 4, 6. If ω_p is less than unity, then ω_0 lies on the extension of the low-frequency segment at $\omega = 1$.

7.2.1.2   Lag (Low-Pass) Filter

If the frequency of the zero becomes very high compared to the loop bandwidth (approximately the frequency where the open-loop gain is unity), it will have little effect on the closed-loop response, except at frequencies near to, or beyond, ω_z. The filter transfer function, Eq. (7.15), now approaches Eq. (7.21) and the damping factor, Eq. (7.14), approaches Eq. (7.24), while the natural frequency, given by Eq. (7.20), does not change. The responses in Eqs. (7.12) and (7.18) are also unaffected. Because 2ζ now equals ω_n/ω_0, Eq. (7.17) can be simplified to Eq. (7.23). The magnitudes of Eqs. (7.18) and (7.23) are plotted in Fig. 3.12, while curves for magnitude and phase, with $N = 1$, are available.[15]

7.2.1.3   Integrator-plus-Lead Filter

When moving the filter pole toward zero, we will no longer maintain unity filter gain at DC. Instead, we will cause the magnitude of the filter response at $\omega = 1$ (or the projection of its -6-dB/octave region through $\omega = 1$) to be

13   Blanchard, pp. 45–138.
14   Gardner, pp. 7–16 and 28–40.
15   Blanchard, pp. 106–110.

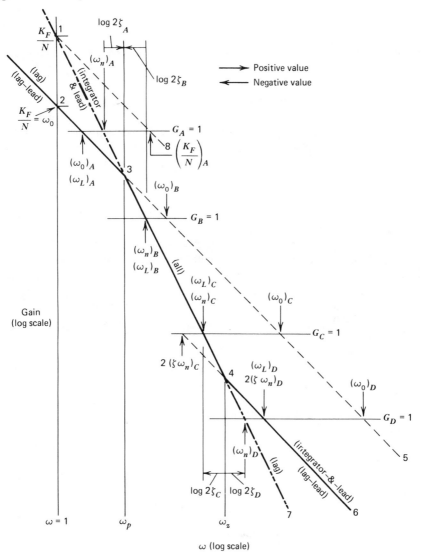

Figure 7.7 Open-loop gain, geometric interpretation (approximate).

unity, by arbitrarily setting ω_{pp} to unity. Thus, Eq. (7.10) becomes

$$\left. \frac{G_{LF}}{K_{LF}} \right|_{\omega_{pp}=1} = \frac{1 + s/\omega_z}{\omega_p + s} . \tag{7.32}$$

As the pole drops toward zero frequency, this approaches Eq. (7.25) in Table 7.1 and ω_0 approaches infinity. The natural frequency from Eq. (7.13) becomes Eq. (7.30) and the damping factor approaches Eq. (7.31). Here K_F and K_F/N equal the total forward and open-loop gains at $\omega = 1$, excepting that, if

ω_z is below 1 rad/sec, they equal the gains using the projection of the filter response below ω_z to $\omega = 1$. Equations (7.11) and (7.12) then lead to Eqs. (7.26)–(7.28). Curves showing these frequency respones are available in the literature.[16]

7.2.1.4 *Geometric Interpretation*

Figure 7-7 shows the open-loop gain and four different unity-gain levels so the effect on ω_n and ζ can be studied as the gain multiplier changes. This figure illustrates many of the important relationships discussed in the previous section and is worthy of detailed study.

The gain for a lag-lead filter is shown as curve 2, 3, 4, 6. For a lag filter, it is at 2, 3, 4, 7. The gain with a loop filter consisting of an integrator and a zero is shown at 1, 4, 6.

Since ω_0 is the loop gain at $\omega = 1$ (assuming the situation shown at 2, 3, 4 in Fig. 7.7, i.e., no integrator and no poles nor zeros between zero and one), it is also the frequency where the gain equals unity after falling at -6 dB/octave. Thus, ω_0 is at the intersection of curve 2, 5 and the unity gain line.

For the lag-lead and lag filters, ω_n is located midway between ω_p and ω_0 on the log plot of Fig. 7.7, since Eq. (7.20) gives

$$\log \omega_n = \tfrac{1}{2}(\log \omega_p + \log \omega_0).$$

For the integrator-lead filter, it is midway between $\omega = 1$ and K_F/N, since Eq. (7.30) gives

$$\log \omega_n = \frac{1}{2} \log \frac{K_F}{N} = \frac{1}{2}\left(\log 1 + \log \frac{K_F}{N}\right).$$

In either case, it follows that ω_n lies at the intersection of line 1, 7 and the unity gain line.

When the loop bandwidth is much closer to ω_p than to ω_z, Eq. (7.14) approaches Eq. (7.24) and we may write

$$\log 2\zeta \approx \log \omega_p - \log \omega_n. \tag{7.33}$$

This is shown near point 3 in Fig. 7.7. Of course, this is exactly true for the lag filter, which has no zero.

When the loop bandwidth is much closer to ω_z than to ω_p, Eq. (7.14) approaches Eq. (7.31) and we write

$$\log 2\zeta \approx \log \omega_n - \log \omega_z. \tag{7.34}$$

This is shown near point 4 in Fig. 7.7. This is exactly correct for the integrator-plus-lead, where there is no nonzero pole.

EXAMPLE 7.4

Problem What is the natural frequency of a second-order loop whose open-loop gain is rolling off at -12 dB/octave at 10 kHz and has a value of 20 dB there?

---

16 Blanchard, pp. 113–116.

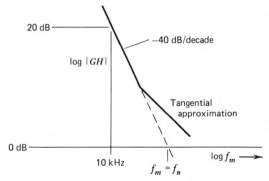

Figure 7.8 Bode plot for Example 7.4.

Solution The natural frequency is at the intersection of the $-12\text{dB}/\text{octave}$ (-40 dB/decade) line and unity gain level. The number of decades between 10 kHz and the unity-gain frequency is

$$\frac{20 \text{ dB}}{40 \text{ dB}/\text{decade}} = \frac{1}{2} \text{ decade}.$$

Therefore, the intersection is at

$$f_n = 10 \text{ kHz} \times 10^{1/2} = 31.6 \text{ kHz}.$$

This is illustrated in Fig. 7.8.

EXAMPLE 7.5

Problem Find the damping factor of a second-order loop with unity open-loop gain at 10 kHz, a filter pole at 5 kHz, and a filter zero at 20 kHz.

Solution The Bode gain plot is shown in Fig. 7.9. From this, the natural frequency is 10 kHz. Using this in Eq. (7.14) gives

$$\zeta = \frac{1}{2}\left(\frac{\omega_p}{\omega_n} + \frac{\omega_n}{\omega_z}\right) = \frac{1}{2}\left(\frac{5 \text{ kHz}}{10 \text{ kHz}} + \frac{10 \text{ kHz}}{20 \text{ kHz}}\right) = \frac{1}{2}.$$

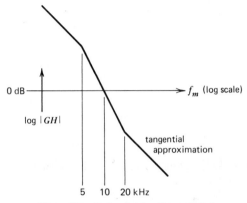

Figure 7.9 Bode plot for Example 7.5.

7.2.1.5   *Circuit Equivalence*[17]

The circuit of Fig. 7.10 has responses, to voltages, which are identical to the responses of the second-order loop to frequency modulation.

EXAMPLE 7.6

Problem   Draw the equivalent response circuit (as in Fig. 7.10) for a synthesizer which has a reference frequency of 100 kHz, an output frequency of 100 MHz, a loop filter pole at 1 kHz, a zero at 10 kHz, and a low-pass filter corner too high to have an important effect.

Solution   It can be seen from Fig. 7.11*a* that the natural frequency is 10 kHz, so the inductor is

$$L = \frac{1}{\omega_n} = \frac{1}{2\pi \times 10^4} = 15.9 \ \mu H.$$

Similarly, the capacitor is

$$C = \frac{1}{\omega_n} = 15.9 \ \mu F.$$

The damping factor is obtained from Eq. (7.14) as

$$\zeta = \frac{1}{2}\left(\frac{\omega_p}{\omega_n} + \frac{\omega_n}{\omega_z} \right) = \frac{1}{2}(0.1 + 1) = 0.55.$$

Total resistance is

$$R_1 + R_2 = 2\zeta = 1.1 \ \Omega.$$

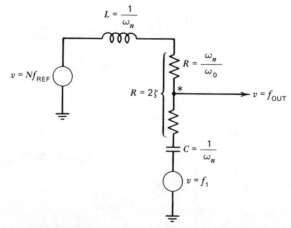

Figure 7.10   Circuit with responses equivalent to second-order loop responses. (*Tap is fully up for integrator plus zero, fully down for low-pass filter.)

17   Blanchard, p. 117.

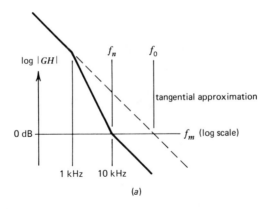

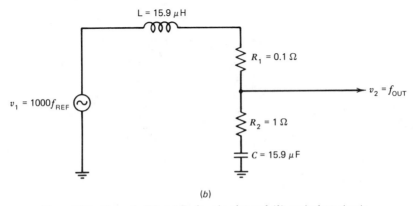

Figure 7.11 Example 7.6: (a) Bode gain plot and (b) equivalent circuit.

The resistance above the tap is

$$R_1 = \frac{\omega_n}{\omega_0} = \frac{2\pi \times 10^4}{2\pi \times 10^5} = 0.1\ \Omega.$$

The voltage source has a multiplier equal to the divide ratio:

$$N = \frac{100\ \text{MHz}}{100\ \text{kHz}} = 1000.$$

The circuit is shown in Fig. 7.11b.

7.2.1.6 Transient Response

The transient response f_{OUT}/f_1 can be easily obtained from formulas and graphs given in the phase-locked-loop literature[18,19] by including $1/N$ in the gain. We will be more interested, however, in f_{OUT}/f_{REF}.

18 Blanchard, pp. 81–101.
19 Gardner, pp. 33–35.

Equations (7.17), (7.23), and (7.27) may be written

$$\frac{f_{\text{OUT}}}{f_{\text{REF}}}(s) = N\omega_n \frac{2\alpha\zeta s + \omega_n}{s^2 + 2\zeta\omega_n s + \omega_n^2}, \tag{7.35}$$

where, for lag,

$$\alpha = 0, \tag{7.36}$$

for integrator plus lead,

$$\alpha = 1, \tag{7.37}$$

and for lag-lead,

$$\alpha = 1 - \frac{\omega_n}{2\zeta\omega_0} = \frac{1}{1 + \omega_z/\omega_0}, \tag{7.38}$$

$$0 < \alpha < 1. \tag{7.39}$$

The corresponding response to a unit step, $U(t)$, at f_{REF} is

$$f_{\text{OUT},R}(t) = N\left\{ U(t) + \frac{1}{2}\left[\left(\zeta\frac{2\alpha - 1}{\sqrt{\zeta^2 - 1}} - 1 \right)e^{-\omega_n t(\zeta - \sqrt{\zeta^2 - 1})} \right.\right.$$

$$\left.\left. - \left(\zeta\frac{2\alpha - 1}{\sqrt{\zeta^2 - 1}} + 1 \right)e^{-\omega_n t(\zeta + \sqrt{\zeta^2 - 1})} \right]\right\} \tag{7.40}$$

$$= N\left\{ U(t) - e^{-\zeta\omega_n t}\left[\cosh\left(\omega_n t\sqrt{\zeta^2 - 1}\right) \right.\right.$$

$$\left.\left. + \zeta\frac{1 - 2\alpha}{\sqrt{\zeta^2 - 1}} \sinh\left(\omega_n t\sqrt{\zeta^2 - 1}\right) \right]\right\}. \tag{7.41}$$

More convenient forms are, for $\zeta = 1$,

$$f_{\text{OUT},R}(t)\big|_{\zeta=1} = N\left\{ U(t) - e^{-\omega_n t}\left[1 + (1 - 2\alpha)\omega_n t \right] \right\} \tag{7.42}$$

and, for $\zeta < 1$,

$$f_{\text{OUT},R}(t)\big|_{\zeta<1} = N\left\{ U(t) - e^{-\zeta\omega_n t}\left[\cos\left(\omega_n t\sqrt{1 - \zeta^2}\right) \right.\right.$$

$$\left.\left. + \zeta\frac{1 - 2\alpha}{\sqrt{1 - \zeta^2}} \sin\left(\omega_n t\sqrt{1 - \zeta^2}\right) \right]\right\}. \tag{7.43}$$

Differentiation of any of these expressions with respect to time reveals that the initial slope of the step response equals $2N\alpha\zeta\omega_n$.

Each of the above responses may be written as an average of the response with a lag filter and with an integrator-plus-lead:

$$f_{\text{OUT, }R}(t)|_\alpha = \alpha f_{\text{OUT, }R}(t)|_{\alpha=1} + (1 - \alpha)f_{\text{OUT, }R}(t)|_{\alpha=0}. \qquad (7.44)$$

This can be shown by use of Eq. (7.41) in Eq. (7.44) and also by observation of the equivalent circuit in Fig. 7.10. This enables the response for various values of α to be obtained as a weighted average of the responses for $\alpha = 0$ and for $\alpha = 1$. These are given in Fig. 7.12. Error signals at larger offsets are shown in the log plot of Fig. 7.13; for the underdamped case, the envelope of the error is plotted. Equation (7.44) does not hold for the envelope.

According to Eq. (3.35), the step response to an excitation at f_1 can be obtained by deleting N and $U(t)$ from Eqs. (7.40)–(7.43) above and changing the sign. In fact, this response, $f_{\text{OUT, }1}(t)$, represents the closed-loop response at any point to a unit step injected there. Equation (7.44) also applies. To summarize:

$$f_{x,x}(t)|_\alpha = f_{x,x}(t)|_{\alpha=1} + (1 - \alpha)f_{x,x}(t)|_{\alpha=0}, \qquad (7.45)$$

$$f_{x,x}(t)|_{\zeta>1} = \left[\exp\left(-\zeta\omega_n t\right)\right]\left[\cosh\left(\omega_n t\sqrt{\zeta^2-1}\,\right)\right.$$

$$\left. + \zeta\frac{1-2\alpha}{\sqrt{\zeta^2-1}}\sinh\left(\omega_n t\sqrt{\zeta^2-1}\,\right)\right] \qquad (7.46)$$

$$f_{x,x}(t)|_{\zeta=1} = \left[\exp(-\zeta\omega_n t)\right]\left[1 + (1-2\alpha)\omega_n t\right] \qquad (7.47)$$

$$f_{x,x}(t)|_{\zeta<1} = \left[\exp(-\zeta\omega_n t)\right]\left[\cos\left(\omega_n t\sqrt{1-\zeta^2}\,\right)\right.$$

$$\left. + \zeta\frac{1-2\alpha}{\sqrt{1-\zeta^2}}\sin\left(\omega_n t\sqrt{1-\zeta^2}\,\right)\right] \qquad (7.48)$$

where $f_{x,x}(t)$ is the closed loop response of frequency at any point x due to a unit frequency step introduced at x.

Note also that the response of phase to a unit phase step is the same as the response of frequency to a unit frequency step.

In many applications, the loop parameters will vary during a transient. This may be due, for example, to the nonlinear character of the VCO tuning curve or to saturation of an operational amplifier. These effects may occur during overshoot, where the operating point of a component is not in a region of normal steady-state operation. The resulting increase or decrease in gain can further accentuate the overshoot that causes the problem. In these cases, it seems likely that the magnitudes of the acquisition time and the overshoot will correspond to some weighted average of the parameters occurring during the acquisition process, so worst-case calculations might be based on parameter extremes occurring during the transient.

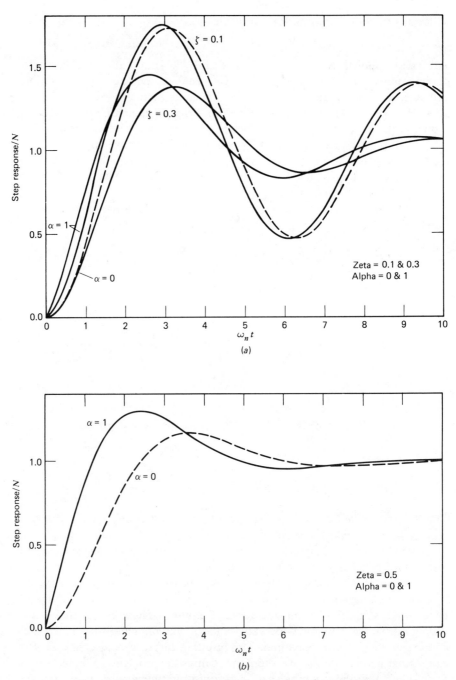

Figure 7.12 Response of output frequency to input frequency step.

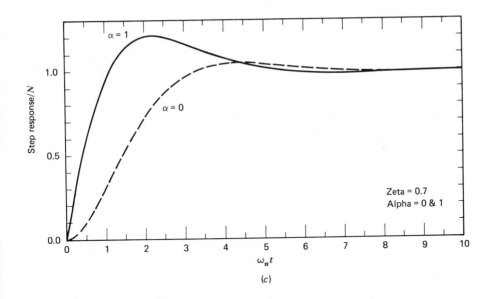

(c)

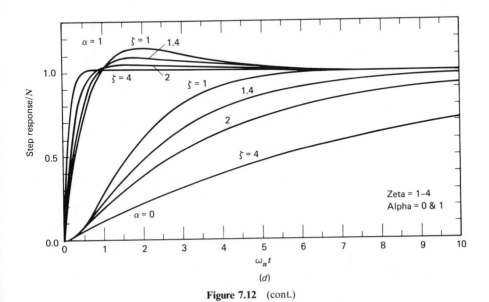

(d)

Figure 7.12 (cont.)

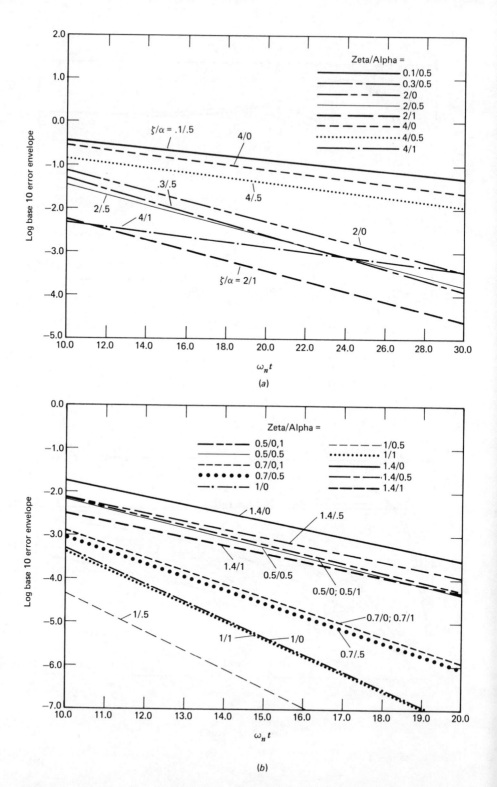

(a)

(b)

7.2.2   Loop With Integrator-plus-Lead-Lag Filter

As has been noted, the response of a loop with an integrator-plus-lead-lag filter can sometimes be approximated by that of a second-order loop by judiciously choosing to neglect one of the singularities in the lead-lag filter. At times, however, this is not satisfactory. Fortunately, the Bode plot is useful in all cases. It indicates stability margins and these imply, qualitatively, the nature of the response, slow and monotonic or fast and ringing. Figure 7.14 illustrates a Bode plot for such a loop. As an aid to plotting accurate gain and phase curves, Figs. 7.15 and 7.16 give gain and phase shift for a lead-lag filter. This information is contained in the impedance of and R-C network which could be used as the feedback network in an op-amp circuit, which would then have gain proportional to that impedance. Note that, for maximum stability margin, unity gain should occur at the geometric mean of the pole and zero frequencies where the excess phase is minimum.

EXAMPLE 7.7

Problem   A type-2 loop has unity gain at 10 kHz. The loop filter has a zero at 1 kHz and a pole at 2 kHz. Find the phase margin. Do not use the tangential approximation.

Solution   From Fig. 7.16,

$$\gamma = \frac{\omega_1}{\omega_p} = \frac{\omega_z}{\omega_p - \omega_z} = 1.$$

At the frequency ω, where the gain is unity,

$$x \triangleq \frac{\omega}{\omega_p} = \frac{10 \text{ kHz} \times 2\pi}{2 \text{ kHz} \times 2\pi} = 5.$$

The margin is the difference between $-90°$ and the filter phase given by Fig. 7.16 for $\gamma = 1$ and $x = 5$, or 5.6°. Alternately, the angle of the filter function,

$$G_{LF} = \frac{j\omega + \omega_z}{j\omega(j\omega + \omega_p)}$$

Figure 7.13   Logarithm of the envelope of the error in output frequency (the difference from the final value) in response to an input frequency step. (For $\zeta \geqslant 1$, the $\log_{10}|\epsilon|$ is plotted, where ϵ is the error. (For $\zeta < 1$, if $x = \omega_n t$,

$$\ln|\epsilon| = -\zeta x + \ln\left|\cos x\sqrt{1 - \zeta^2} + \zeta\frac{1 - 2\alpha}{\sqrt{1 - \zeta^2}} \sin x\sqrt{1 - \zeta^2}\right|$$

$$= -\zeta x + \ln\left|\left[1 + \frac{(1 - 2\alpha)^2}{1/\zeta^2 - 1}\right]^{1/2} \sin\left(\theta + x\sqrt{1 - \zeta^2}\right)\right|.$$

The plotted envelope is

$$\frac{1}{\ln 10}\left(-\zeta x + \frac{1}{2}\ln\left|1 + \frac{(1 - 2\alpha)^2}{1/\zeta^2 - 1}\right|\right).$$

This is equal to or greater than $\log_{10}|\epsilon|$.)

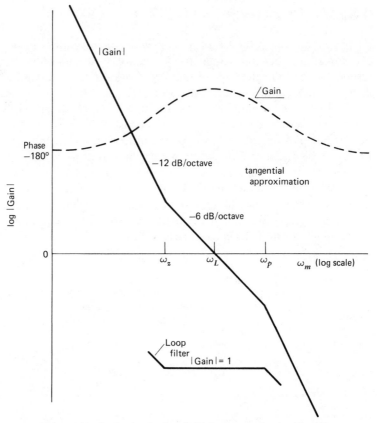

Figure 7.14 Bode plot for loop with integrator-plus-lead-lag filter.

is

$$-90° + \tan^{-1}\left(\frac{\omega}{\omega_z}\right) - \tan^{-1}\left(\frac{\omega}{\omega_p}\right) = -90° + \tan^{-1}10 - \tan^{-1}5 = 5.6° - 90°,$$

giving a margin of 5.6°.

Figure 7.17 shows the effect of this type of loop on the phase-noise characteristics of a typical VCO and on a typical reference-phase-noise profile, where the pole and zero are well separated. The closed-loop noise plots are obtained from the open-loop transfer function GH by

$$\frac{f_{\text{OUT}}}{f_1} = \frac{1}{1 + GH} \tag{7.49}$$

and

$$\frac{f_{\text{OUT}}}{f_{\text{REF}}} = \frac{G}{1 + GH}. \tag{7.50}$$

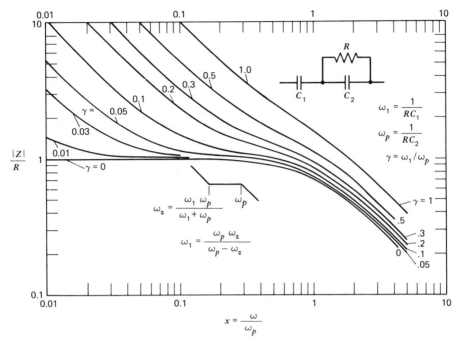

Figure 7.15 Magnitude of Z, integrator-plus-lead-lag.

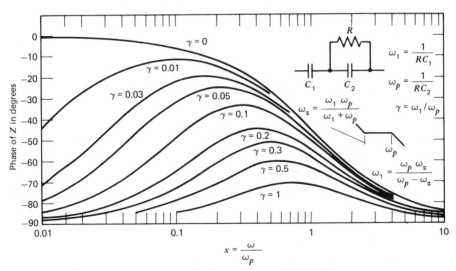

Figure 7.16 Phase of Z, integrator-plus-lead-lag.

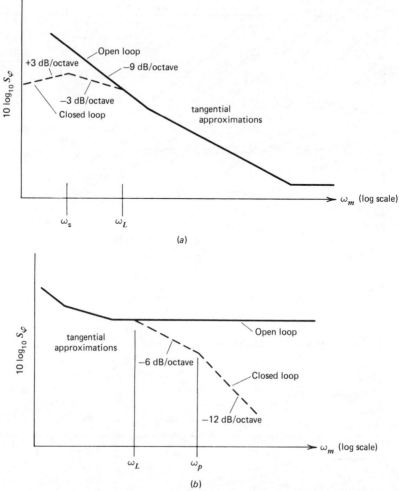

Figure 7.17   Open- and closed-loop phase noise with an integrator-plus-lead-lag filter (tangential plots): (*a*) phase noise of VCO and (*b*) phase noise of reference at synthesizer output.

Well below ω_L, in Fig. 7.14, GH is large and Eqs. (7.49) and (7.50) become

$$\frac{f_{\mathrm{OUT}}}{f_1}(\omega \ll \omega_L) \approx \frac{1}{GH} \tag{7.51}$$

and

$$\frac{f_{\mathrm{OUT}}}{f_{\mathrm{REF}}}(\omega \ll \omega_L) \approx \frac{G}{GH} = \frac{1}{H} = N. \tag{7.52}$$

Therefore, the open-loop VCO noise (f_1), is attenuated by the open-loop gain, but the reference noise is multiplied by N.

Well above ω_L, GH is much smaller than one and Eqs. (7.49) and (7.50) become

$$\frac{f_{\text{OUT}}}{f_1}(\omega \gg \omega_L) \approx 1 \tag{7.53}$$

and

$$\frac{f_{\text{OUT}}}{f_{\text{REF}}}(\omega \gg \omega_L) \approx G. \tag{7.54}$$

Therefore, well above ω_L, the open-loop VCO noise is unaffected by the loop and the reference noise is multiplied by the forward gain. At ω_L, $|GH| = 1$, so $|G| = 1/H$ or N, so the high- and low-frequency approximations meet at ω_L.

This procedure for obtaining closed-loop frequency response from open-loop response is not peculiar to the type-2 loop, but is applicable to any loop for which the open-loop characteristics are known. The root-locus method may also be used.[20,21]

EXAMPLE 7.8

Problem A loop filter has an integrator and a zero at 10 kHz with a pole at 400 kHz. Compute the reduction in VCO phase noise, at 10 Hz from the spectral center, due to the loop, if the open-loop gain is unity at 20 kHz.

Solution The attenuation equals the open-loop gain at 10 Hz. The open-loop gain is shown in Fig. 7.18, from which can be seen that the value at 10 Hz is

$$|GH(10\text{ Hz})| = 20\log_{10}\!\left(\frac{20\text{ kHz}}{10\text{ kHz}}\right) + 40\log_{10}\!\left(\frac{10\text{ kHz}}{10\text{ Hz}}\right)$$

$$= 126\text{ dB}.$$

Other effects that may have to be accounted for in $GH(\omega)$ are frequency responses of the phase detector (Section 5.2), IF filters (Section 7.1.4), the VCO or ICO tuning input network, and circuit delays.

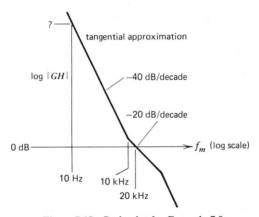

Figure 7.18 Bode plot for Example 7.8.

20 Weaver.
21 Truxal.

7.2.3   Similarities Between Responses

We have alluded to similarities between the performances of loops which possess similar gain and phase profiles near ω_L. Here we will explain this, based on the assumptions that open-loop gain is very high at frequencies well below ω_L, as in Fig. 7.5. These assumptions lead to Eq. (7.51)–(7.54).

Well below ω_L the response to f_{REF} [Eq. (7.52)] is almost independent of the actual magnitude and phase of G, so singularities in this region make little difference. Well above ω_L the response to f_{REF} [Eq. (7.54)] is very small and almost equals the foward gain. Thus the loop is essentially open there and computation of the frequency response is relatively simple. Since little energy is transferred at these frequencies, we do not expect the response in this region to have a significant influence on the gross transient response. Thus the largest components of the frequency and transient responses depend primarily upon the open-loop gain and phase profiles near ω_L, to which Eq. (7.50) is quite sensitive. Similarly, the frequency response to f_1 depends almost entirely on open-loop gain, and is quite small, well below ω_L [Eq. (7.51)] and is almost independent of the open-loop gain and phase well above ω_L [Eq. (7.53)] so that neither of these regions strongly influences the frequency response nor the gross transient response. Again, these responses depend mainly on the open-loop gain and phase near ω_L where Eq. (7.49) is highly influenced by them. Therefore, if two loops have the same ω_L and the same poles and zeros in the vicinity of ω_L, their transient respones will look similar and the frequency responses at those frequencies which are not highly attenuated will be similar. Still, there can be significant differences in small quantities: a very-high-frequency pole will make very-high-frequency responses to f_{REF} even weaker and changes in very-low-frequency gain will influence the transient error (relative to final value) as it becomes very small.

7.3   THE DIGIPHASE® SYNTHESIZER

An alternate means for generating a voltage proportional to phase difference is illustrated in Fig. 7.19. An accumulator increases its stored number N_A by the value of the control word C each cycle of the reference frequency. The contents of the accumulator represents the desired number of phase cycles of the synthesizer output at some point in the reference cycle. A free-running divider, or frequency counter, counts the synthesizer output frequency. The contents of this counter N_C are compared to the accumulator contents to determine phase error. Phase error may be converted to voltage by switching a current i_p into the loop filter whenever the accumulator contents exceeds the counter contents, or by employing some other type of phase detector to compare the beginning and end of such a pulse. At steady state, the output frequency is

$$F_{\text{OUT}} = CF_{\text{REF}}. \qquad (7.55)$$

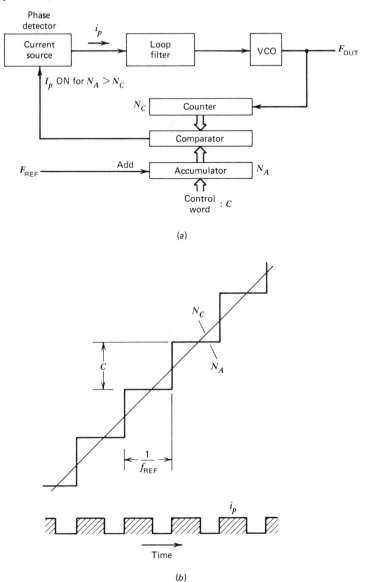

Figure 7.19 Synthesizer with count comparator: (*a*) block diagram and (*b*) N_C, N_A, and i_p.

As described so far, this method is probably more complex than previously described mechanizations and it does not seem to offer any advantages. However, by building on this technique, a synthesizer can be created which is capable of resolution much finer than its reference frequency, and we have seen how low reference frequencies force narrow bandwidths with their accompanying disadvantages.

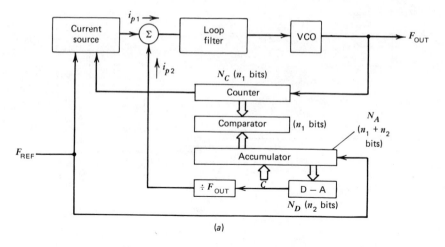

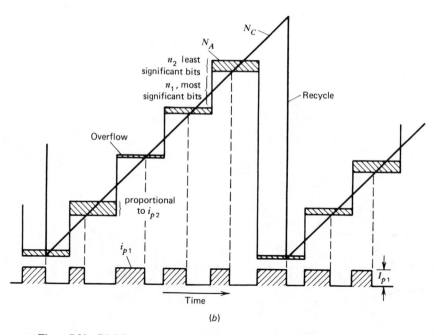

Figure 7.20   Digiphase technique: (*a*) block diagram and (*b*) N_C, N_A, and i_p.

The apparently impossible task of obtaining resolution smaller than reference frequency is accomplished by a technique called Digiphase[22,23]. In this
technique, n_2 bits (assuming a binary mechanization for discussion), representing frequency increments less than F_{REF}, are added to the control word and to
the accumulator. The accumulator is then longer than the comparator and the

22   Gillette.
23   Braymer.

counter, as indicated in Fig. 7.20a. Even though N_2, the number comprising the n_2 least significant bits of the accumulator, is not compared to N_C, it nevertheless enters into the determination of the average output frequency by overflowing into the accumlator's n_1 most significant bits (N_1). This is illustrated in Fig. 7.20b. The output frequency is still given by Eq. (7.55) so, at steady state, N_A and N_C are equal at the same phase during each cycle. However, unless f_{OUT} is an integer multiple of F_{REF}, N_2 increases each reference period, excepting at overflow, and, therefore, the rest of N_A, the part being compared to N_C, achieves equality with N_C earlier each period, until overflow. This is illustrated in Fig. 7.20b, where the current pulse i_{p1} becomes narrower and narrower until overflow, at which time it suddenly widens. The illustration is for a control number where N_2 is about one-third of the least significant bit of N_1. If N_2 were almost equal to 2^{n_2} the picture would be considerably different but, in any case, the average value of i_{p1} will cause F_{OUT} to be produced.

Note that F_{REF} enters the current-source block in Fig. 7.20 but not in Fig. 7.19. In Fig. 7.20, the current pulse starts at the beginning of each reference period. This is the same as the criterion used in Fig. 7.19, $N_A > N_C$, except when N_A reaches capacity and recycles. Since this must happen regularly with any real counter and accumulator, it is necessary that i_{p1} start at the beginning of each reference period, as in Fig. 7.20.

So far, we still do not have a very effective system because the variations in i_{p1} can occur at a frequency as low as the synthesizer's resolution. Filtering these could force the loop to be about as narrow as with standard techniques for the same resolution. This problem is solved by adding a current i_{p2}, which exactly compensates for the change in i_{p1}. We will now determine what form i_{p2} must take.

The difference in width between two adjacent current pulses equals the difference between the time to count N_1 cycles of F_{OUT} and the reference period:

$$\Delta t_{p1} = \frac{N_1}{F_{OUT}} - \frac{1}{F_{REF}} \tag{7.56}$$

$$= \frac{1}{F_{OUT}}\left(N_1 - \frac{F_{OUT}}{F_{REF}}\right) \tag{7.57}$$

$$= \frac{1}{F_{OUT}}\left[N_1 - \left(N_1 + \frac{N_2}{2^{n_2}}\right)\right] \tag{7.58}$$

$$= -\frac{N_2}{2^{n_2}F_{OUT}} . \tag{7.59}$$

The change in charge flowing during the pulse is this change in pulse width multiplied by the peak pulse current,

$$\Delta Q_{p1} = -\frac{I_{p1}N_2}{2^{n_2}F_{OUT}} . \tag{7.60}$$

This change in charge can be compensated by gating a current,

$$I_{p2} = \frac{I_{p1}N_2}{N_x 2^{n_2}}, \qquad (7.61)$$

where N_x is a constant, into the loop filter for N_x counts of F_{OUT}, thus producing a charge

$$Q_{p2} = I_{p2}t_{p2} = \frac{I_{p1}N_2}{N_x 2^{n_2}} \frac{N_x}{F_{OUT}} = -\Delta Q_{p1}. \qquad (7.62)$$

Alternately, a current equal to

$$I'_{p2} = \frac{I_{p1}}{2^{n_2}} \qquad (7.63)$$

may be gated into the loop filter for N_2 counts of F_{OUT}. One reported[22] mechanization uses the decimal equivalent of Eq. (7.63) separately for each decade of N_2.

The amount of FM produced by this system depends on the accuracy with which i_{p2} is generated. The modulating frequencies are a function of N_2 and will, therefore, change as different output frequencies are programmed.[24]

A synthesizer has been reported[22,25] using this technique which operates from 40 to 51 MHz with a reference frequency of 100 kHz. It has a resolution of 1 Hz and near-in spurious sidebands typically less than -70 dBc.

The loop diagram for analysis is the same as Fig. 2.8b except the output of the phase detector may be considered to be a current (as it could for the charge-pump phase detector). For the Digiphase system, however, N is not necessarily a whole number, since the ratio of f_{OUT} to f_{REF} can contain decimal or binary fractions. The phase detector gain is

$$K_{\varphi c} = I_{p1}. \qquad (7.64)$$

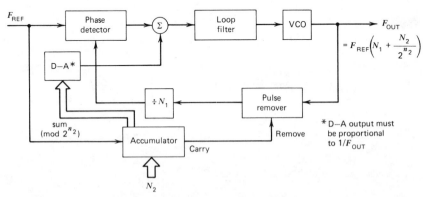

Figure 7.21 Fractional-N synthesizer.

24 Gibbs.
25 Dana Series 7000 Digiphase®.

Another reported method, called "fractional-N" synthesis,[24] is basically the same as the Digiphase method but, in detail, is more like the more usual phase-locked synthesis. It is represented by Fig. 7.21. Here only the fractional part (N_2) of the divide number is accumulated and, when a carry occurs, a pulse is "removed" from the input to the N_1 divider, thus effectively increasing the count by one for that period.

A further modification of Digiphase reduces the low-frequency content of the phase-detector output by means of digital circuitry, thus easing the requirements for accuracy in compensating for the phase-detector output waveform.[26] The contents of the accumulator in Fig. 7.21 are integrated in a second accumlator of the same capacity. Each time the second accumulator overflows, a pulse is removed from N_1 and then a pulse is added on the next count. This increases the phase for one reference period, counteracting the steadily decreasing phase due to F_{OUT}/N_1 being higher than F_{REF} (Fig. 7.22).

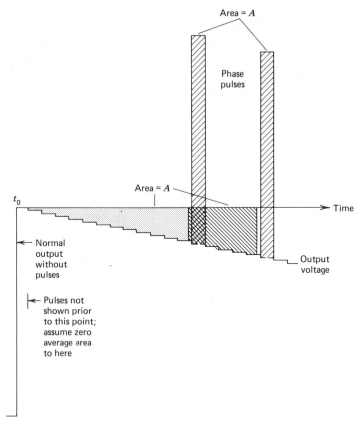

Figure 7.22 Phase-detector output with zero running-average area.

26 RACAL. Patents applied for.

The time between the leading edges of these pulses of phase change is such that the sum of the sequential values in the first accumulator has totaled to the accumulator capacity, equivalent to a change of one in N_1. This implies that the area under the phase-detector waveform (referenced to its initial value and neglecting the phase pulses) between leading edges of phase pulses has the same magnitude as the phase pulse. Thus a running average of zero is maintained. This is effective in reducing the low frequencies produced when F_{OUT} is near a multiple of F_{REF}, when many such corrections are made during each period of the phase-detector output, but not when F_{OUT} is near midway between two multiples of F_{REF}. However, the lowest modulation frequency produced then is approximately $\frac{1}{2} F_{REF}$ and this component is easier to filter.

PROBLEMS

7.1   In Fig. 7.1, what is the maximum output frequency range if the synthesizer components have the following parameters and an increase in output frequency causes an increase in divider output frequency?
$F_{REF} = 100$ kHz;
$N = 900$ to 1900;
low-pass filter cutoff, 230 MHz;
band-pass filter passband, 650 to 1080 MHz;
F_{M2}, selectable, 900 MHz, 1000 MHz, 1100 MHz;
F_{M1}, selectable, 3000 or 3300 MHz.

7.2   In Fig. 7.1, the signal strength at F_{M1} is $+20$ dBm and the first mixer has 30-dB LO (F_{M1}) to RF (other input) isolation. The coupler is a 10-dB coupler with 20 dB of directivity.

  (a)   The VCO output circuit contains an isolator that provides a good output match. How much signal at F_{M1} appears in the output?

  (b)   If the isolator is removed and the VCO then reflects $\frac{1}{10}$ of the incident signal voltage, what is the maximum signal level at F_{M1} which could appear at the output?

  (c)   In (b), what is the minimum?

7.3   A synthesizer has a 3-kHz loop bandwidth and a 20-kHz sample frequency. The input to the frequency divider is FM modulated at 41 kHz with a peak deviation of 500 Hz. What is the predominant sideband produced on the output? Give its amplitude and offset from the desired output.

7.4   An interfering signal at 100 MHz enters a divider input along with a 10-MHz signal to be divided.

  (a)   What is the theoretical absolute maximum level for the interfering signal so it will not cause a miscount if the desired signal is 1-V peak and a maximum of 0.4V peak is needed in the absence of interference?

(b) If the interfering signal frequency is at 100 kHz, what is its maximum level?

7.5 In Fig. 7.1, the loop gain at 1 rad/sec is 60,000. The low-pass filter (FL3) is flat to 100 kHz, and the bandpass filter is a single-pole LC filter, centered at the frequency passing through it (10 MHz), with a 40-kHz 3-dB bandwidth. If the loop filter (FL1) is a simple low-pass, what is its corner frequency to give 45° phase margin?

7.6 What is the natural frequency of a second-order loop whose open-loop gain is rolling off at −40 dB/decade at 30 kHz and has a value of 10 dB there?

7.7 What are the natural frequency and damping factor of a second-order loop with unity open-loop gain at 10 kHz, a filter pole at 1 kHz and a filter zero at 5 kHz?

7.8 A type-2 second-order loop has the same natural frequency as a type-1 loop with a lag-lead filter. K_F/N is 5 times greater for the type-2 loop than for the type-1 loop. What is the pole frequency for the type-1 loop filter? Hint: use Fig. 7.7.

7.9 Draw the equivalent response circuit (as in Fig. 7.10) for a synthesizer which has a reference frequency of 10 kHz, an output frequency of 100 MHz, a loop filter pole at 10 kHz, a zero at 20 kHz, unity loop gain at 20 kHz, and a low-pass filter corner too high to have an important effect.

7.10 A loop filter has an integrator with a zero at 1 Hz and a pole at 4 kHz.

(a) If the open-loop gain is unity at 4 Hz, how long will it take for the signal to reach 90% of its final value after a frequency step?

(b) How long will be required if the open-loop gain is unity at 1 kHz? (Hint: approximate the loop with a second-order loop.)

7.11 A type-2 loop has unity gain at 4 kHz. The loop filter has a zero at 500 Hz and a pole at 2 kHz. What is the phase margin? Do not use the tangential approximation.

7.12 Using the Bode-plot tangential-gain approximation, compute the reduction in VCO phase noise at 100 Hz due to the loop in Problem 7.10(b).

7.13 Draw the control-system (mathematical) diagram for a Digiphase synthesizer with the following parameters:
VCO tuning sensitivity, 1 MHz/V;
Loop filter, 1-kΩ resistor in parallel with a 0.16-μF capacitor to ground;

Peak current from phase detector into filter (I_{p1}), 1mA;
Output frequency, 37,526,204 Hz;
Reference frequency, 100 kHz.

7.14 The frequency synthesized by a Digiphase synthesizer is

$$NF_{REF} \pm \delta f$$

where

$$\delta f \ll F_{REF}.$$

and N is an integer. Show that the ratio of desired output voltage to the sideband voltage level at $\pm \delta f$ is the same as the attenuation of the component of the phase-detector output voltage at δf (that is its reduction due to cancellation by the D-A), assuming the attenuation is large.

8

Large-Signal
Performance

An understanding of the large-signal performance of the phase-locked synthesizer is important in determining whether a particular loop will ever achieve lock and under what range of initial conditions this will occur.

The phase-locked synthesizer could be categorized as a nonlinear sample-data system with signal-dependent sample rate. As such, it is resistant to exact analysis. Therefore, we approximate it by systems which are analyzable, in order to understand and predict its performance. These various approximations are outlined in Fig. 8.1. So far, we have approximated it by linear systems, an appropriate approximation for small deviations from steady state. We have, in addition, assumed constant sampling frequency, also an appropriate approximation for most small-signal responses, once steady state has been achieved. And, in some cases, we can ignore sampling altogether. Now we shall consider the process by which steady state is achieved and here we generally cannot use a linear approximation. A number of results are available, however, based on nonlinear analysis, simulation, and observation.

We will first consider the performance of a simple loop, acquiring phase lock, by writing difference equations describing the changes during each sample period. The analysis will not employ the constant-sampling-rate approximation, a fact which leads to some surprising results.

Then, as we move to nonlinear systems, we drop consideration of sampling, as we list some available equations relating to acquisition of lock and then consider the effects of component offsets and saturation on acquisition. Next we shall examine the origin of the signal which causes lock to eventually occur when cycles are being skipped, and this will lead us to the discovery of a type of false lock.

We shall also describe a second type of false lock occurring in the acutal nonlinear, signal-dependent-sampling case. And, we shall discuss how com-

	Sampling		
	None	Constant rate	Signal dependent
Linear	4, 7	5	8.1
Non linear	8.2 8.3 8.4		8.5 8.6 8.7

Figure 8.1 Varying degrees of approximation to the actual loop. Numbers show sections employing each approximation extensively.

puter simulation can be used to predict nonlinear performance and how loops can be tested for acquisition.

8.1 SIMPLE LOOP ACQUIRING LOCK

We now consider the performance of the simple loop in Fig. 8.2, one without a loop filter, as it acquires phase lock. The held phase error (i.e., the phase difference relative to steady state) and the divider output frequency during the nth period are called φ_n and F_{sn}. The initial frequency is F_{s0}, the result of an initial phase error, φ_0. A frequency error will cause a change in phase error at the end of the nth period which is given, in cycles, by

$$\Delta\varphi_{nc} = F_{REF}\Delta T_n, \tag{8.1}$$

where ΔT is the error in the period, T_s, of F_s. ΔT_n may be written in terms of the frequency during the preceding period:

$$\Delta\varphi_{nc} = F_{REF}\left(\frac{1}{F_{s(n-1)}} - \frac{1}{F_{REF}}\right) = -\frac{\Delta f_{(n-1)}}{F_{s(n-1)}}, \tag{8.2}$$

where

$$\Delta f_n \underset{\Delta}{=} F_{sn} - F_{REF}. \tag{8.3}$$

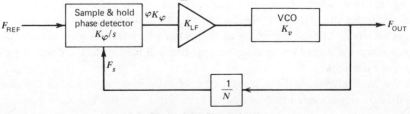

Figure 8.2 Simple loop.

8.1.1 Optimized Gain Shaping

The frequency error will be corrected at the first sample after it is detected if the phase change is

$$\Delta\varphi_1 = -\varphi_0. \tag{8.4}$$

By Eq. (8.2), this means

$$\varphi_{0c} = \frac{\Delta f_0}{F_{s0}}. \tag{8.5}$$

This implies a tuning characteristic such that

$$F_s \varphi_c = \Delta f \underline{\Delta} F_s - F_{\text{REF}} \tag{8.6}$$

or

$$F_s(\varphi) = \frac{F_{\text{REF}}}{1 - \varphi_c}, \tag{8.7}$$

resulting in an output frequency,

$$F_{\text{OUT}}(\varphi) = NF_s(\varphi) = \frac{F_{ss}}{1 - \varphi_c} \tag{8.8}$$

where

$$F_{ss} \underline{\Delta} NF_{\text{REF}}. \tag{8.9}$$

Equation (8.8) defines the optimum forward gain, from φ_c to F_{OUT}. Unfortunately, it defines a different tuning curve for each steady-state frequency F_{ss}, since by Eq. (8.8) φ_c is zero when $F_{\text{OUT}} = F_{ss}$ for all F_{ss}. Since the general case involves multiple-output frequencies, this cannot be accomplished, so we concentrate on satisfying Eq. (8.7) only when the operating point is near final value. Then φ_c is small and Eq. (8.8) may be written

$$F_{\text{OUT}}(\varphi) \approx F_{ss} e^{\varphi_c}. \tag{8.10}$$

This equation merely states that the output frequency is the exponential of the phase. It can be satisfied for all F_{ss} because, if it holds for some value F_{ss}, we can write, for any other value,

$$F_{\text{OUT}}(\varphi) \underline{\Delta} F'_{ss} e^{\varphi'_c}, \tag{8.11}$$

where

$$\phi'_c \underline{\Delta} \varphi_c - \ln(F'_{ss}/F_{ss}). \tag{8.12}$$

Note that the optimum loop gain may be obtained from Eq. (8.10) as

$$\bar{\omega}_0 \equiv \frac{1}{N} \frac{dF_{\text{OUT}}}{d\varphi_c}\bigg|_{F_{\text{OUT}} = F_{ss}} = \frac{F_{ss}}{N} = F_{\text{REF}} \tag{8.13}$$

$$= \frac{1}{T_s}, \tag{8.14}$$

corresponding to the center of the unit circle in the Z plane (Fig. 5.37). Thus the forward gain compensates for the changing value of N and constant loop gain is produced.

We define the relative gain as

$$G_R \triangleq \frac{\omega_0}{F_{REF}} \tag{8.15}$$

and the realtive gain error as

$$E_G \triangleq G_R - 1. \tag{8.16}$$

We define the relative error in sample frequency at the nth period as

$$E_n \triangleq \frac{F_{sn} - F_{REF}}{F_{REF}} . \tag{8.17}$$

There are two effects that slow the reduction of an initial error. One is the error E_G in achieving the optimum value given by Eq. (8.13) and the other is the inexactness of that optimum gain, since Eq. (8.10) is only an approximation for small frequency errors. Appendix 8A shows the following regarding these effects.

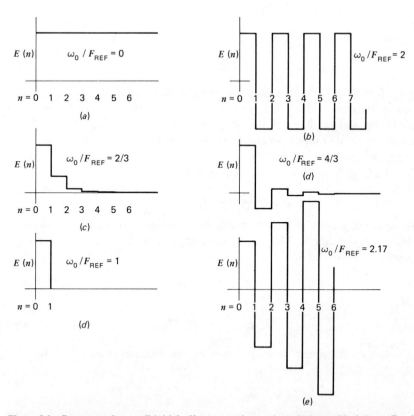

Figure 8.3 Responses for small initial offsets at various gains relative to ideal ($\omega_0 = F_{REF}$).

If the initial relative frequency error E_0 is small (compared to E_G, and to 1 and $1/G$), then the error in the nth period due to gain inaccuracy is

$$E_n \approx (-E_G)^n E_0. \tag{8.18}$$

Some responses plotted according to this equation are shown in Fig. 8.3. The required number of corrections (periods) to attain $\pm E_n$ is

$$n = \frac{\log_b |E_n/E_0|}{\log_b |E_G|}, \tag{8.19}$$

where the logarithmic base b is arbitrary.

If the relative gain error E_G is zero, and E_0 is small compared to one, then the nth error, due to approximation of Eq. (8.8) by Eq. (8.10) is

$$E_n \approx 2 \left(\frac{E_0}{2} \right)^{2^n} \tag{8.20}$$

and the number of periods to achieve $\pm E_n$ is

$$n = \log_2 \frac{\log_b (E_n/2)}{\log_b (E_0/2)}. \tag{8.21}$$

8.1.2 Constant Gain

We have shown how the loop reacts when the changes are small enough so the exponential gain can be considered constant [Eq. (8.18)]. Now, however, we wish to carry the analysis further to see what happens when the gain *is* constant, but the changes are not necessarily small.

We will investigate the effect of two sequential corrections. With constant gain, the frequency change is given, in terms of the sampled phase change, as

$$F_{sn} - F_{s(n-1)} = \omega_0 \Delta \varphi_{nc} \tag{8.22}$$

and this may be combined with Eq. (8.2) to give

$$F_{sn} - F_{s(n-1)} = \omega_0 \left(\frac{F_{REF}}{F_{s(n-1)}} - 1 \right). \tag{8.23}$$

Combining this with Eq. (8.17), we obtain

$$E_n = E_{n-1} - \frac{\omega_0}{F_{REF}} \left(1 - \frac{1}{E_{n-1} + 1} \right) \tag{8.24}$$

$$= E_{n-1} \left(1 - \frac{G_R}{1 + E_{n-1}} \right). \tag{8.25}$$

Iterating this last equation gives the exact expression for relative error as a function of the relative error two periods earlier:

$$E_n = E_{n-2}\left(1 - \frac{G_R}{1 + E_{n-2}}\right)\left[1 - \frac{G_R}{1 + E_{n-2}\left[1 - \dfrac{G_R}{(1 + E_{n-2})}\right]}\right]. \quad (8.26)$$

If E_{n-2} were small, we would obtain

$$E_n \approx E_{n-2}(1 - G_R)^2, \qquad (8.27)$$

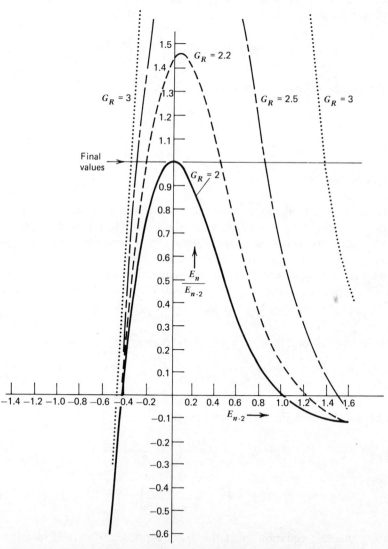

Figure 8.4 Ratio of relative error after two samples versus initial relative error.

showing that no change occurs if G_R is 0 or 2, as indicated in Figures 8.3a, 8.3b, and 5.37. If G_R is 2, a constant amplitude oscillation would be sustained, as predicted by Z-transform theory.

However, if we plot the ratio E_n/E_{n-2} from Eq. (8.26) as a function of E_{n-2}, as in Fig. 8.4, we find some very interesting results. By starting at various arbitrary points in Fig. 8.4 and following the sequences of operating points, we can see that final values lie along the intersection of the curves given by Eq. (8.26) and a line where $E_n = E_{n-2}$. The abscissas at these intersections give the positive and negative deviations from zero error for constant amplitude oscillations at the given gains. We find that an oscillation will only be sustained, with a relative gain (G_R) of 2, at zero amplitude $(E_{n-2} = 0)$. More gain is required to sustain a finite amplitude oscillation; the larger the amplitude, the more gain is required. Such an effect does not appear in a Z-transform analysis because the sampling rate is not signal dependent there, but Eq. (8.26) represents exactly the performance of the assumed loop.

EXAMPLE 8.1

Problem Loop gain is 2.2 times F_{REF} and the initial frequency is high by 80% of the "steady-state" value (NF_{REF}). Find the amplitude of oscillations after they have reached a constant level.

Solution The realtive gain G_R is 2.2 and the initial relative error is 0.8. Using Eq. (8.26) for accuracy, but following the results on Fig. 8.4, we obtain the following information. Going up from $E_{n-2} = 0.8$ to the curve for $G_R = 2.2$, we find that $E_2/E_0 = 0.372$, so $E_2 = 0.298$. Repeating this process, we find subsequent alternate values of E_n to be 0.367, 0.410, 0.426, 0.431, and 0.431. Thus eventually every other value of E_n will be 0.431. The other "final value" at this gain is -0.232, meaning that E_n alternates between -0.232 and 0.431.

8.2 FORMULAS FOR NONLINEAR BEHAVIOR

We now present various equations describing the nonlinear performance of phase-locked loops for various phase-detector characteristics and open-loop gain functions. None of these will describe the type-2 loop with a lead-lag filter and none take the effects of finite sampling rates into account. However, as before, we must use what information is available, judiciously applying it where appropriate, that is, where the sampling frequency and the low-pass corner frequency, if the loop is type 2, are high enough to have little effect within the loop bandwidth.

First, we will describe the process of acquisition of lock in general, in order to define the terms to be used. We will refer to frequencies at the divider output because our equations will then correspond to the literature for loops without dividers. One of these frequencies is the VCO "mistuning" divided by N, that is,

$$\delta\omega_{MIS} \underset{=}{\Delta} \Omega_{s0} - \Omega_{REF}, \tag{8.28}$$

where Ω_{s0} is the value of the divider output frequency, Ω_s, that occurs in the absence of a reference signal. Without a reference input, the phase-detector output contains a component at Ω_s plus a "DC" component. The component at Ω_s is highly attenuated by the loop filter while the "dc" component tunes Ω_s to the center frequency Ω_{s0}. Because the theoretical cases studied have symmetrical phase detectors with no limiting following, the "dc" component equals the phase-detector voltage in the center of its range. Because of this symmetry, the ranges may be given as $\pm$ deviations of Ω_{s0} about Ω_{REF}, or of Ω_{REF} about Ω_{s0}.

A locked loop will maintain lock over the hold-in range, $\pm\omega_H$. This is simply the range of $\delta\omega_{MIS}$ corresponding to the phase detector's output range. Thus, once lock has occurred, Ω_{s0} may drift from Ω_{REF} by $\pm\omega_H$ or Ω_{REF} may be tuned slowly from Ω_{s0} by $\pm\omega_H$ without a loss of lock.

No wider than the hold-in-range is the pull-in, or frequency acquisition, range, $\pm\omega_{PI}$. If $\delta\omega_{MIS}$ comes within the pull-in range, the loop will eventually lock, although it may skip many cycles in the process. The time for the cycle skipping to cease is called the pull-in or frequency-acquisition time T_{PI}.

No wider than the pull-in range is the lock-in, or seize frequency, range, $\pm\omega_S$. Within this range of $\delta\omega_{MIS}$, cycle skipping will not occur. The corresponding time is the lock or seize time T_S but, since the loop pulls in exponentially, this is not meaningful unless the final frequency or phase error is specified.

Pull-in may occur beyond $\pm\omega_{PI}$ and the loop may lock without cycle skipping beyond $\pm\omega_S$, depending on initial conditions. These are values for the worst initial conditions (e.g., initial phase).

In order to summarize, let us consider what occurs if the reference frequency is connected to the loop after the oscillator has been free running at $N\Omega_{s0}$. If $|\delta\omega_{MIS}|$ is greater than ω_{PI}, the loop may* skip cycles indefinitely. If $|\delta\omega_{MIS}|$ is between ω_{PI} and ω_S, the loop may* skip cycles for as long as T_{PI}. It will then cease cycle skipping and pull into lock. If $|\delta\omega_{MIS}|$ is less than ω_S, lock

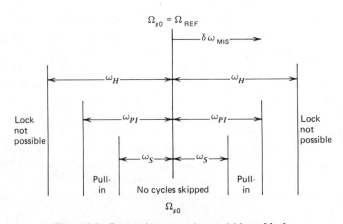

Figure 8.5 Ranges important in acquisition of lock.

*Depending on initial phase.

will occur without cycle skipping. Once the loop has locked, if the reference frequency is slowly changed, lock will be broken when $|\delta\omega_{\text{MIS}}|$ reaches ω_H. The divider output will then move back toward Ω_{s0}, although there will be some frequency modulation due to the beat note at the phase-detector output. These ranges are illustrated in Fig. 8.5.

8.2.1   First-Order Loop

The simple first-order loop not only serves as an approximation for some practical synthesizer loops, but also behaves in some ways like the higher-order loops to be described.

This loop has a constant 90° open-loop phase shift and behaves like a low-pass circuit, as shown in Fig. 3.7c, regardless of gain. Since there will, therefore, be no overshoot, any operating point within the hold-in range can be acquired without cycle skipping. Therefore, for the three phase-detector types, we have

$$\pm\omega_{PI} = \pm\omega_S = \pm\omega_H = \pm\omega_0 \times \begin{cases} \pi & \text{for sawtooth type} & (8.29\text{A}) \\ \pi/2 & \text{for triangular type} & (8.29\text{B}) \\ 1 & \text{for sine type}^\dagger & (8.29\text{C}) \end{cases}$$

for the first-order loop. The times of interest are those required for a simple exponential to get from initial to final value,

$$T = \frac{1}{\omega_0} \ln \frac{\Delta\omega_f}{\Delta\omega_i}, \tag{8.30}$$

where $\Delta\omega_i$ and $\Delta\omega_f$ are the initial and final errors (i.e., offsets from final value), respectively. Since, with a sinusoidal phase detector, ω_0 changes during pull-in, an effective value, something between the initial and final values, should be employed.

EXAMPLE 8.2

Problem   Two 10-MHz signals are locked together using a sinusoidal phase detector. The loop filter is a low-pass which has 3-dB attenuation at 10 kHz. Lock can be maintained as the reference frequency varies from 10.000, 4 MHz to 9.998, 4 MHz. How much time is required to come within 1 Hz of final frequency after an initial 200-Hz offset?

Solution   We approximate this as a first-order loop because the effect of the filter will be small at the loop bandwidth, ω_0, which, by Eq. (8.29C), would equal ω_H if the operating point were in the center of the hold-in range. Since it is not, we have

$$\omega_0 < \omega_H = 2\pi \frac{(10.000,4 - 9.998,4)}{2} \times 10^6$$
$$= 2\pi \times 1000 \ll 2\pi \times 10,000,$$

confirming our assumption.

†Where ω_0 is taken at the middle of the phase-detector range.

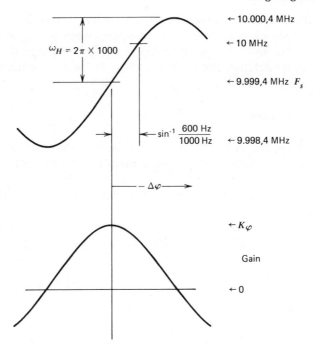

Figure 8.6 Phase-detector characteristic for Example 8.2.

Since most of the transient will occur near final frequency, we will use the gain there. Therefore, as illustrated by Fig. 8.6, we multiply ω_H by the ratio of K_φ at the operating point to K_φ at the center of the sinusoidal characteristic:

$$\omega_0 = \omega_H \cos\left(\sin^{-1}\frac{600\,\mathrm{Hz}}{1000\,\mathrm{Hz}}\right) = 2\pi \times 800.$$

The required time will then be given by Eq. (8.30) as

$$T \approx \frac{1}{2\pi \times 800}\ln\left(\frac{200\,\mathrm{Hz}}{1\,\mathrm{Hz}}\right) = 1.1\,\mathrm{msec}.$$

8.2.2 Second-Order Loop

The hold-in range is still given by Eq. (8.29).

The lock-in ranges for the three phase-detector types are

$$\omega_S = \frac{\omega_n^2}{\omega_z} \times \begin{cases} \pi & \text{for sawtooth type*} & (8.31A) \\ \pi/2 & \text{for triangular type*} & (8.31B) \\ 1 & \text{for sine type if } \omega_P, \omega_z \gg \omega_L.^\dagger & (8.31C) \end{cases}$$

*ω_S is the largest initial error frequency that can exist, when the initial phase is at the edge of the phase-detector characteristic, without causing the phase to go out of the linear range. For triangular and sawtooth characteristics, this can be obtained by adding the derivative of the phase error (at the phase-detector input), due to a maximum phase offset, to the frequency error, due to an initial frequency offset of ω_S, and setting the sum to zero at zero time.[3] The initial rate of phase

8.2.2.1 Sinusoidal Phase Detector

If $\omega_0 \gg \omega_z$, the pull-in range is[4, 5]

$$\omega_{PI} \approx 2\omega_0(x - x^2)^{\frac{1}{2}}, \tag{8.32}$$

where $x = \omega_p/2\omega_z$. For a low-pass filter ($\omega_z \to \infty$), ω_{PI} is the smaller of the hold-in range (ω_0) and[6]

$$\omega_{PI} \approx 3\zeta\omega_0\left[(0.423 + 1.19\zeta^4)^{\frac{1}{2}} - 1.092\zeta^2\right]^{\frac{1}{2}}. \tag{8.33}$$

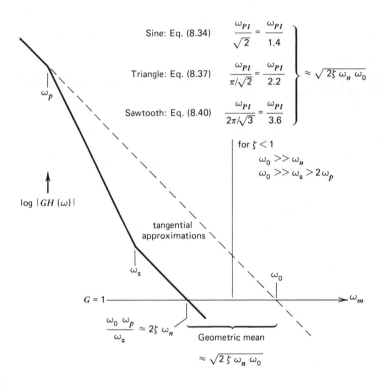

Figure 8.7 Pull-in related to open-loop gain plot. ω_{PI} varies from 1.4 to 3.6 times the geometric mean (average on a log plot) of $2\zeta\omega_n$ and ω_0. (See restrictions in text.)

change is then zero and any greater offset would cause the response to move initially away from final value, and thus over the nonlinearity.

[†]If ω_p and ω_z are small compared to ω_L, then, near ω_L, the loop acts like a first-order loop with the same gain. Therefore, $\omega_S = \omega_L$, which was ω_0 for the first-order loop, but is given by Eq. (8.31C) for this case.[1, 2]

1 Gardner, pp. 43–46.
2 Kroupa, p. 177
3 Byrne, pp. 587 and 588.

4 Greenstein.
5 Rey.
6 Richman, "Color Carrier . . . ", p. 125.

If $\omega_0 \gg \omega_z$ and also $\omega_n \ll 2\zeta\omega_0$,

$$\omega_{PI} \to 2(\zeta\omega_0\omega_n)^{\frac{1}{2}}. \tag{8.34}$$

This is shown relative to the Bode plot in Fig. 8.7. If $\delta\omega_{MIS}$ is neither too close to ω_{PI} nor to ω_S, the pull-in time is[1,6]

$$T_{PI} \approx \frac{(\delta\omega_{MIS})^2}{2\zeta\omega_n^3} \text{ if } \omega_0 \gg \omega_z. \tag{8.35}$$

The maximum lock-in time for a 10:1 reduction in frequency error has been given as[7]

$$T_S \lesssim \frac{10}{\omega_0} \frac{\omega_z}{\omega_p}. \tag{8.36}$$

EXAMPLE 8.3

Problem   A loop with a sinusoidal phase detector and a lag-lead filter has the following parameters:

$$N = 100$$
$$K_F = 10^6$$
$$\omega_z = 250,$$
$$\omega_p = 25.$$

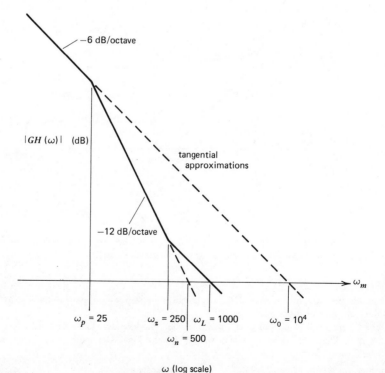

Figure 8.8   Bode plot for Example 8.3.

7   Robson.

Estimate the pull-in frequency, seize frequency, pull-in time from 500-Hz error, and seize time.

Solution The Bode gain plot is shown in Fig. 8.8. The damping factor (ζ) is given, by Eq. (7.14), as one. Since $\omega_0 \gg \omega_z$ and $\omega_n \ll 2\zeta\omega_0$, Eq. (8.34) applies giving

$$\omega_{PI} \approx 4472.$$

Since $\omega_L = 4\omega_z$, we will use Eq. 8.31(C) for the seize frequency, although we might suspect its accuracy:

$$\omega_s \approx 1000.$$

The initial error is

$$\delta\omega_{MIS} = 2\pi \times 500 = 3142,$$

which is, hopefully, not too close to either ω_{PI} nor ω_s, so that Eq. (8.35) applies, giving

$$T_{PI} \approx 39.5 \text{ msec.}$$

The lock-in, or seize, time for a 10:1 frequency error reduction is given by Eq. (8.36) as

$$T_S \lesssim 10 \text{ msec.}$$

8.2.2.2 Triangular Phase Detector

With a triangular phase-detector characteristic, the pull-in range has been reported as[8]

$$\omega_{PI} \approx \pi\sqrt{\zeta\omega_0\omega_n} \quad \text{if } \omega_n \ll \omega_0 \text{ and } \zeta < 1. \tag{8.37}$$

This is shown in Fig. 8.7. It bears the same ratio to Eq. (8.34) as do the hold-in ranges for the two phase-detector types (sine and triangle). For $\zeta = 0.5$, Eq. (8.37) is about 10% high when $\omega_n = 0.05\omega_0$ but it is 32% high when $\omega_n = 0.1\omega_0$.

For a low-pass filter, if $\zeta \geqslant 1$, the pull-in range equals the hold-in range ($\frac{1}{2}\pi\omega_0$), otherwise[9]

$$\omega_{PI} = \frac{\pi}{2}\omega_0\frac{1-x}{1+x}, \tag{8.38}$$

where*

$$x = \frac{(1+\zeta^2)^{\frac{1}{2}} - \zeta^2}{(1+\zeta^2)^{\frac{1}{2}} + \zeta^2}\exp\left[\frac{-\pi + \tan^{-1}(1/\zeta^4 - 1)^{\frac{1}{2}}}{(1/\zeta^2 - 1)^{\frac{1}{2}}}\right]. \tag{8.39}$$

Since the gain is constant over the phase-detector range, the response equations, Eqs. (7.40)–(7.43), can be used to compute lock-in time.

8.2.2.3 Sawtooth Phase Detector

Acquisition with a sawtooth phase-detector characteristic is of particular interest in phase-locked synthesizers. This problem has been studied by

---

*Use $0 < \tan^{-1}(1/\zeta^4 - 1)^{\frac{1}{2}} \leqslant \pi/2$.

8 Cahn.
9 Protonotarios.

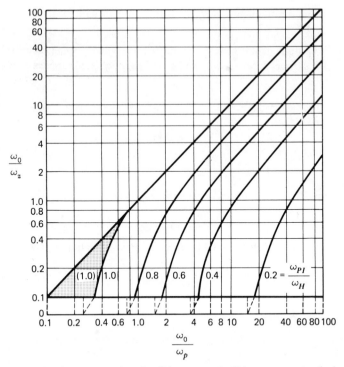

Figure 8.9 Contours of constant ratios of pull-in range to hold-in range, sawtooth characteristic. From Byrne, Fig. 13, p. 584. Copyright 1962, The American Telephone and Telegraph Company. Reprinted by permission.

Goldstein[10] and his results are incorporated in a design-oriented companion paper by Byrne[11], which is a rather thorough discussion of the mathematical properties of this type of loop.

Figure 8.9 shows the pull-in range relative to the hold-in range, Eq. (8.29A). The pull-in frequency may be written as

$$\omega_{PI} \approx 2\pi\omega_0 \sqrt{\frac{\omega_p}{3\omega_z}} \approx 2\pi\sqrt{\tfrac{2}{3}}\ \sqrt{\zeta\omega_0\omega_n}\ , \tag{8.40}$$

if $\omega_0 \gg \omega_z > 2\omega_p$.

This is shown in Fig. 8.7. For a low-pass filter, it is

$$\omega_{PI} = \pi\omega_0 \tanh \frac{\pi}{2(1/\zeta^2 - 1)^{\frac{1}{2}}} \quad \text{if } \zeta \leqslant 1 \tag{8.41}$$

and

$$\omega_{PI} = \pi\omega_0 \quad \text{if } \zeta \geqslant 1. \tag{8.42}$$

10 Goldstein.
11 Byrne.

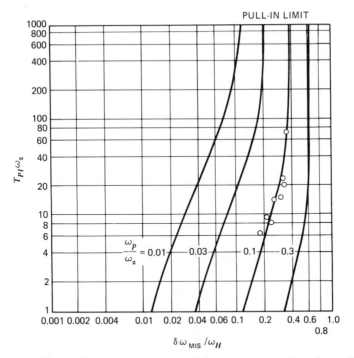

Figure 8.10 Pull-in time for a sawtooth characteristic when $\omega_z \ll \omega_0$. From Byrne, Fig. 17, p. 589. Copyright 1962, The American Telephone and Telegraph Company. Reprinted by permission.

The pull-in time is given by Fig. 8.10 and the lock-in time can be computed from Eqs. (7.40)–(7.43).

EXAMPLE 8.4

Problem Compute the pull-in frequency for the loop in Example 8.3, except use a triangular and a sawtooth phase detector. Compare the results with the sinusoidal case. Also compare the pull-in time of the loop using the sawtooth phase detector to that using the sinusoidal phase detector.

Solution For the traingular case, Eq. (8.37) gives*

$$\omega_{PI} = 7025,$$

which is larger by $\pi/2$ than with a sinusoidal phase detector. For the sawtooth characteristic, since

$$\omega_z \ll \omega_0 \text{ and } \omega_z > 2\omega_p,$$

the pull-in frequency may be obtained from Eq. (8.40) as

$$\omega_{PI} = 11,470.$$

This is more than 1.6 times what was obtained for the triangular characteristic and 2.6 times the range for a sinusoidal phase detector.

*Equation (8.37) was derived for $\zeta < 1$ but we expect no drastic change as $\zeta \to 1$.

The seize frequency for the sawtooth case is given by Eq.(8.31A) as 500 Hz, so cycle skipping will (barely) be eliminated. The pull-in time with a sinusoidal phase detector was about 40 msec, giving

$$T_{PI}\omega_z = 10.$$

In Fig. 8.10, this value on the curve for

$$\frac{\omega_p}{\omega_z} = 0.1$$

corresponds to an initial frequency error of

$$\delta\omega_{MIS} = 0.24\omega_H = 7540 \rightarrow 1200 \text{ Hz.}$$

Thus, with a sawtooth phase detector, pull-in from 1200-Hz initial error can occur in the same 40 msec that was required for pull-in from 500 Hz with a sinusoidal phase detector.

8.2.2.4   General Phase-Detector Characteristic

For a phase detector with any odd-symmetrical characteristic equal to

$$g(\varphi) = -g(-\varphi) = g(\varphi + 2\pi) \tag{8.43}$$

and unity gain at zero (where ω_0 is computed),

$$\left. \frac{dg(\varphi)}{d\varphi} \right|_{\varphi=0} = 1, \tag{8.44}$$

the pull-in frequency has been given as

$$\omega_{PI} = \omega_0 \left\{ 2 \left[\langle g^2 \rangle \left(\frac{\omega_p}{\omega_z} \right) + \langle \varphi g \rangle \left(\frac{\omega_p}{\omega_0} \right) + \left(\frac{\pi \omega_p}{\omega_0} \right)^2 \right]^{\frac{1}{2}} - \frac{2\pi\omega_p}{\omega_0} \right\}, \tag{8.45}$$

where $\langle g^2 \rangle$ is the average value of the square of the characteristic,

$$\langle g^2 \rangle = \frac{1}{\pi} \int_0^\pi g^2(\varphi) \, d\varphi, \tag{8.46}$$

and $\langle \varphi g \rangle$ is the average of the product φg,

$$\langle \varphi g \rangle = \frac{1}{\pi} \int_0^\pi \varphi g(\varphi) \, d\varphi. \tag{8.47}$$

Formulas for computing T_{PI} are given in the same paper.[12]

8.2.2.5   The Effects of Component Offsets and Saturation

By Eqs. (8.34), (8.37), and (8.40), we can see that if we include a perfect integrator in the loop filter, the pull-in range is theoretically unlimited. However, there is more than the absence of a perfect integrator to prevent us from actually achieving unlimited pull-in range. Figure 8.11 shows a loop with an offset voltage at the phase-detector output. This may be due to the phase

12 Mengali.

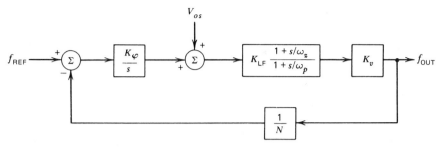

Figure 8.11 Loop with offset.

detector or to the operational amplifier in the loop filter. The offset may be small, but it will exist.

The reason that the loop will theoretically acquire lock over an infinite range is that any finite correction voltage can tune the oscillator an arbitrary amount if it is multiplied by the infinite dc gain of an ideal integrator. However, the offset voltage is added to the correction voltage before amplification and the correction voltage must overcome it. If the correction voltage is not larger than the offset, the sum may not have the correct polarity and acquisition will then be impossible.

Even if suffcent correction voltage could be generated to overcome the offset, it might be prevented from doing so by saturation of components, in particular the operational amplifier and the VCO (tuning response). If the output frequency does not respond to changes in phase-detector output, the process that generates the correction voltage during cycle skipping is defeated. Therefore, when such a voltage is required, the loop-filter gain should be low enough that the following components will not saturate due to the offset (as well as other influences: temperature, etc). This does not imply that pull-in cannot occur when a loop variable swings through a nonlinearity during the process, but this can certainly make analysis difficult. For the remainder of the discussion, we assume that component saturation will be prevented. We then seek the design parameters to give the best pull-in range. We assume a sawtooth characteristic and that the pull-in frequency is given by Eq. (8.40).

The voltage offset, $\pm V_{os}$, is equivalent to a frequency offset of

$$\pm \omega_{os} = \pm V_{os} K_{LF} K_{vr}. \tag{8.48}$$

The net pull-in range is the pull-in range in the absence of offset less the frequency offset (assuming its sign is unknown). At the synthesizer output, this is $\pm \Delta \omega_{OUT}$, where

$$\Delta \omega_{OUT} = N \omega_{PI} - \omega_{os} \tag{8.49}$$

$$= 2\pi N \omega_0 \left(\sqrt{\frac{1}{3X}} - \varphi_{oc} \right), \tag{8.50}$$

where φ_{oc} is the phase equivalent, in cycles, of V_{os},

$$\varphi_{oc} = \frac{V_{os}}{K_{\varphi c}} , \qquad (8.51)$$

and

$$X \triangleq \frac{\omega_z}{\omega_p} . \qquad (8.52)$$

If the hold-in range is to be held constant, then X must be made small to optimize $\Delta\omega_{OUT}$, suggesting that a first-order loop is required. In that case, we would use Eq. (8.29A) in Eq. (8.49) and obtain

$$\Delta\omega_{OUT} = 2\pi N\omega_0\left(\frac{1}{2} - \varphi_{oc}\right) \qquad (8.53)$$

and we find that lock would occur as long as the offset is smaller than the maximum phase-detector output.

If, however, the loop bandwidth is to be held constant, then, returning to our assumption of a lag-lead filter, we shall find the configuration to maximize the pull-in range for a given value of ω_L. Assuming that ω_p and ω_z are well within ω_L, ω_0 is approximately $\omega_L X$ and Eq. (8.50) can be written as

$$\frac{\Delta\omega_{OUT}}{\omega_L} = 2\pi N\left(\sqrt{\frac{X}{3}} - X\varphi_{oc}\right). \qquad (8.54)$$

Setting the derivative with respect to X equal to zero, we obtain the value for X for maximum pull-in range as

$$X = \frac{1}{12\varphi_{oc}^2} , \qquad (8.55)$$

Thus, in a practical circuit, high low-frequency gain can actually reduce the pull-in range for a given loop bandwidth. With this optimum ratio of ω_z to ω_p, the pull-in range is

$$\Delta\omega_{OUT} = \frac{\pi N\omega_L}{6\varphi_{oc}} . \qquad (8.56)$$

8.3 THE PULL-IN PROCESS

We shall now consider what causes a loop, which is skipping cycles, to eventually pull into lock (if it does). The procedure to be followed is outlined in Fig. 8.12. Initially, the frequency at the divider output, Ω_x, is assumed to have a steady value of Ω_x'. This is an approximation because, as we will subsequently show, there is FM at this point and because Ω_x may also be slowly changing. Ω_x' at the phase detector input results in a difference frequency, $\delta'\omega$, at its output, where

$$\delta'\omega \triangleq \Omega_x' - \Omega_{REF}. \qquad (8.57)$$

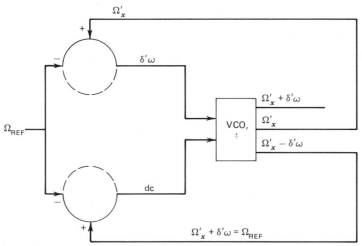

Figure 8.12 Derivation of frequency correction voltage during pull-in.

This, in turn, modulates the VCO, producing FM sidebands which enter the phase detector. One of these sidebands results in a "DC" component at the phase-detector output and this is of a magnitude that will cause the divider output to be

$$\Omega''_x = \omega_D + \Omega_{x0}, \tag{8.58}$$

where Ω_{x0} is the center frequency, which occurs with zero phase-detector output, and ω_D is the offset produced by the "dc" component. If Ω''_x equals the originally assumed value, Ω'_x, then the assumption is verified. If Ω''_x is higher than Ω'_x, then the value of Ω_x will increase, and conversely. By studying the regions where Ω_x tends to increase and decrease, we can gain a better understanding of the pull-in process.

For a sinusoidal phase detector, the output due to Ω'_x mixing with Ω_{REF} will be

$$\text{Re}\, v_p = K_{\varphi r}\, \text{Re}\, e^{j\delta'\omega t}. \tag{8.59}$$

The transfer function from the phase-detector output to the divider output equals the open-loop transfer function divided by the phase detector's transfer function:

$$G'_r(\delta\omega) = \frac{GH(\delta\omega)}{K_{\varphi r}/(j\delta\omega)}. \tag{8.60}$$

Multiplying this by v_p, we obtain the frequency deviation at the divider output.

$$\Delta'\omega(t) = \text{Re}\left[j\delta'\omega GH(\delta'\omega)e^{j\delta'\omega t}\right] \tag{8.61}$$

$$= \delta'\omega |GH(\delta'\omega)| \cos\left[\delta'\omega t + \underline{/GH(\delta'\omega)} + \tfrac{1}{2}\pi\right]. \tag{8.62}$$

The corresponding phase deviation is given by

$$\Delta'\varphi_r(t) = \int \Delta'\omega(t)\,dt = |GH(\delta'\omega)|\sin\left[\delta'\omega t + \underline{/GH(\delta'\omega)} + \tfrac{1}{2}\pi\right]. \quad (8.63)$$

Assuming that the modulation index β at the divider output is small, the signal there becomes[13]

$$v_D \approx A_D\left(\cos\Omega'_x t - \tfrac{1}{2}\,\beta\left\{\cos\left[(\Omega'_x - \delta'\omega)t - \underline{/GH(\delta'\omega)} - \tfrac{1}{2}\pi\right]\right.\right.$$
$$\left.\left. - \cos\left[(\Omega'_x + \delta'\omega)t + \underline{/GH(\delta'\omega)} + \tfrac{1}{2}\pi\right]\right\}\right). \quad (8.64)$$

Substituting the peak phase deviation from Eq. (8.63) for β and employing Eq. (8.57), we obtain

$$v_D \approx A_D\left\{\cos\Omega'_x t - \tfrac{1}{2}|GH(\delta'\omega)|\cos\left[\Omega_{\mathrm{REF}}t - \underline{/GH(\delta'\omega)} - \tfrac{1}{2}\pi\right]\right.$$
$$\left. + \tfrac{1}{2}|GH(\delta'\omega)|\cos\left[2\Omega'_x t - \Omega_{\mathrm{REF}}t + \underline{/GH(\delta'\omega)} + \tfrac{1}{2}\pi\right]\right\}. \quad (8.65)$$

The second component in Eq. (8.65) is at the reference frequency and will, therefore, cause a "dc" output from the phase detector. The magnitude will be reduced relative to normal phase-detector output by the relative level of this sideband and by any deviation from phase quadrature. Taking these factors into account, the "dc" voltage produced at the phase-detector output is

$$v_{\mathrm{dc}} = K_{\varphi r}\tfrac{1}{2}|GH(\delta'\omega)|\sin\underline{/GH(\delta'\omega)} \equiv \tfrac{1}{2}K_{\varphi r}\,\mathsf{Im}\left[GH(\delta'\omega)\right]. \quad (8.66)$$

The resulting frequency offset at the phase-detector input is obtained by multiplying by the "dc" open-loop gain, excepting the phase-detector gain,

$$\omega_D = v_{\mathrm{dc}}\frac{\omega_0}{K_{\varphi r}} = \frac{\omega_0}{2}\,\mathsf{Im}\left[GH(\delta'\omega)\right]. \quad (8.67)$$

This is sketched as curve I in Fig. 8.13 for a typical open-loop transfer function.

Also shown, as curve II, is the steady-state value of ω_D as given by Eq. (8.58) for $\Omega'_x = \Omega''_x$. When curve I is above curve II, at some value of Ω'_x, the "dc" voltage generated is sufficient to produce a frequency higher than Ω'_x, so Ω_x will increase, and conversely. Therefore, point S can be seen to be a point of stable equilibrium, whereas point U is an unstable equilibrium point. If the initial frequency offset ($\delta'\omega$) is greater than that at point U, $\delta\omega$ will settle at point S and a true lock will not be achieved. If the initial offset is less than that at point U, it will decrease to a locked condition.

As Ω_{x0} moves further above Ω_{REF} (F moves right), the "stable" point, S, occurs where Ω'_x almost equals Ω_{x0} and ω_D is very weak; that is, the loop is essentially open.

As the separation between Ω_{x0} and Ω_{REF} decreases, curve II (moving left) no longer intersects curve I and phase lock appears certain. Ω_{x0} is then within the pull-in range. At any value of Ω'_x, ω_D is less than what is required to maintain that separation from Ω_{x0}, and the loop begins to pull in, faster and

13   Westman, pp. 21-6–21-8.

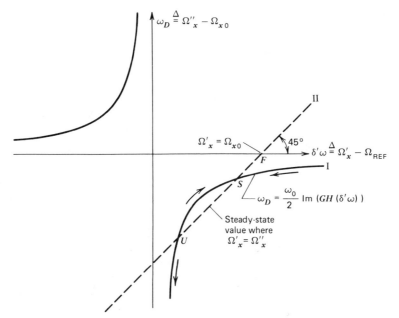

Figure 8.13 Pull-in characteristic.

faster as the separation between the curves at the actual value of Ω_x increases. As the final frequency is approached, the frequency modulation also becomes more pronounced. The phase-detector output can be seen to develop the required "dc" signal by spending more time at phases which produce the required output polarity[14-16]. This can be seen in Fig. 8.20.

Consider now the influence of loop gain. As ω_0 decreases, point S approaches point F, that is, steady-state Ω_x approaches Ω_{x0}; the loop has lost influence on the VCO. But, as ω_0 increases, the curves eventually fail to intersect, and pull-in occurs, unless curve I crosses $\omega_D = 0$. That would imply a phase shift, at the crossing point, of

$$\underline{/GH(\delta'\omega)} = 180°. \tag{8.68}$$

Under these conditions, high gain can *cause* the curves to intersect.

Many of these same regions and effects are evident on phase-plane plots such as Viterbi's,[17] although those plots are valid in regions where our present assumptions fail. Note in particular that, as the operating point on curve I approaches the ordinate axis for phase lock, the modulation index will tend to grow to the point where our small modulation-index assumption fails. At the same time, the approximation of constant Ω_x will become poor.

---

14 Gardner, p. 44, Fig. 4-9.   **16** Klapper, p. 101, Fig. 5-12(a).
15 Kroupa, p. 176, Fig. 6-11.   **17** Viterbi, pp. 60 and 62, Figs. 3.13–3.17.

8.4   FALSE LOCK DUE TO PHASE SHIFT[18, 19]

Figure 8.14 is also a plot of Eq. (8.67) for a hypothetical loop transfer
function. It differs from Fig. 8.13 in that the latter is monotonic, whereas this
second plot shows the effect of considerable phase shift in $GH(\delta\omega)$ with
increasing frequency. (Also, the ordinate axis is normalized.) It is typical in the
sense that the phase lag continues to increase with frequency offset as
amplitude drops.

Initially, assume a perfect integrator in the loop filter. Then the curve of
steady-state ω_D/ω_0 (curve II) lies along the axis of abscissas. Note the
additional stable points at $\pm\omega_y$. Any frequency error beyond $\pm\omega_{yy}$ will tend
to result in a false lock near $\pm\omega_y$, where the phase of GH is $\pm360°$, or at a
higher offset near some higher multiple of $\pm360°$. With a perfect integrator in
the loop filter, any correction voltage will cause the operating point to move
and operation at some point on the $\delta'\omega$ axis will result. With finite gain, the
steady-state operation point will be offset from ω_y sufficiently to develop the
bias required for tuning to that point and an intersection with curve II will not
occur at higher values of $\delta'\omega$.

The frequency offset at which false lock can occur (ω_{FL}) is shown at S in
Fig. 8.14. From Eq. (8.67), this occurs at

$$\omega_D = \omega_{FL} - \omega_{x0} = \tfrac{1}{2}\omega_0 \, \mathrm{Im}\big[\, GH(\omega_{FL})\,\big], \qquad (8.69)$$

where

$$\omega_{x0} \triangleq \Omega_{x0} - \Omega_{REF}, \qquad (8.70)$$

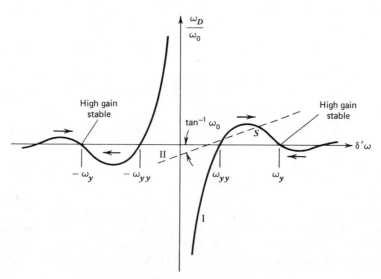

Figure 8.14   Pull-in with significant phase shift.

18   Gardner, pp. 84–90.
19   Develet.

so the required value of ω_0 is

$$\omega_0 = \frac{2(\omega_{FL} - \omega_{x0})}{\mathsf{Im}\left[GH(\omega_{FL})\right]}. \tag{8.71}$$

In some ways the "out-of-lock" condition represented by Fig. 8.13 is similar to the "false lock" of Fig. 8.14; there is a continuum of characteristics between them but their extreme forms act differently. In the low-gain, "out-of-lock" case, changing the reference frequency has little effect on the VCO, which is essentially free-running, and the VCO frequency can be easily changed by retuning. In the high-gain, "false-lock", the difference frequency, $\delta\omega$, is held near the value that gives 360° phase shift (or a multiple thereof), so the oscillator will tend to track any change in Ω_{REF} and will resist retuning. Moreover, the "out-of-lock" condition can be cured by increasing loop gain whereas the "false lock" is remedied by decreasing gain, although, in some cases, changing gain may just turn one condition into the other.

One particularly troublesome situation occurs when the loop employs an integrator and sweeps the VCO center frequency to achieve lock. Since any correction voltage is sufficient to eventually cause lock, it may be necessary to sweep fast enough so the sweep cannot be stopped[20] by the false-lock voltage, or to use a quadrature phase detector to detect a locked condition to stop the sweep.

The potential for false lock is increased by a multiplicity of poles which cause rapid phase shift in $GH(\delta\omega)$. For this reason, the use of multipole IF filters tends to aggrevate the situation. However, real systems have many additional poles at high modulating frequencies. For example, at high enough frequencies, the op-amp will introduce many poles. The fact that its gain is low will reduce the size of the peaks in Fig. 8.14, but they will still be there. Although this type of false lock is not often encountered in synthesizer design, one should be aware of the potential, especially when the frequencies of undesired poles are not orders-of-magnitude higher than the loop bandwidth.

EXAMPLE 8.5

Problem A loop is used to phase lock a microwave oscillator to a reference frequency derived by multiplying the output of a crystal oscillator. Under some conditions, the VCO appears to lock 2 MHz above the reference frequency. The free-running VCO frequency (center of hold-in range) is 0.9 MHz above the reference frequency and the loop bandwidth is 100 kHz. At 2 MHz, the loop filter has dropped 12 dB ($\frac{1}{4}$) in gain from the value it had at ω_L and it has a phase shift of 220°. What would be the value of ω_0 necessary for a false lock?

Solution Since the loop gain was unity at ω_L, at 2 MHz the imaginary part is

$$\mathsf{Im}\,GH(2\ \text{MHz}) = \frac{100\ \text{kHz}}{2\ \text{MHz}}\left(\tfrac{1}{4}\right)\sin[2\pi(-220° - 90°)]$$

$$= 9.6 \times 10^{-3}.$$

20 Gardner, pp. 47, 52, 53.

Using Eq. 8.71, we obtain

$$\omega_0 = \frac{2(2 \text{ MHz} - 0.9 \text{ MHz})2\pi}{9.6 \times 10^{-3}}$$

$$= 1.4 \times 10^9.$$

8.5 FALSE LOCK DUE TO SAMPLING

If the sampling rate is very fast, a frequency offset will cause the phase-detector characteristic to be traced out, by the phase-detector output voltage, with the period of the difference frequency. This is illustrated in Fig. 8.15*b*. Also apparent, however, is the slightly granular nature of the output. The average value of the output can be affected by this granularity. If Fig. 8.15*b* were part of a repetitive picture, each cycle of output looking the same, then a shift of the sample pulse train to the left would cause an identical decrease in output voltage for each sample and for the average value. This is shown by the characteristic of Fig. 8.15*a*. A very-narrow-range average-voltage-versus-phase characteristic exists, centered in the middle of the phase detector's output range. If the loop bandwidth is small compared to the difference frequency, and if the difference frequency can be attained within the reduced-range characteristic shown in Fig. 8.15*a*, then a false lock is possible. The signal at the beat frequency can be ineffective, due to the low gain at that frequency, but the voltage, averaged over many beat cycles, can form an error signal to

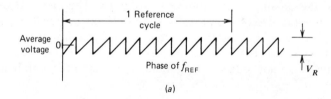

(a)

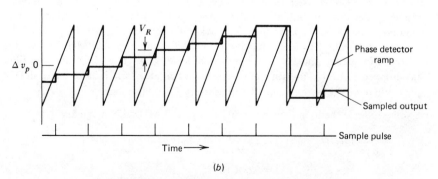

(b)

Figure 8.15 Sampled output when $F_s \neq F_{REF}$.

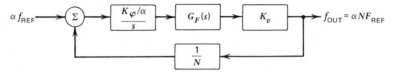

Figure 8.16 Loop for $\alpha : 1$ false lock.

lock the loop. For these conditions the loop in false lock can be represented by Fig. 8.16. The effective reference frequency is higher by the ratio of false-lock to correct-lock frequencies, α. The phase-detector gain constant is reduced by α because, although the ratio of output-voltage deviation to sample-time deviation does not change, the deviation in the sample time represents more phase deviation at the higher divider-output frequency.

When α is close to unity, there are many samples per cycle of error frequency and we expect performance like that of a continuous system. Under these circumstances, the hold-in range for the false lock is very small and the beat note out of the phase detector is at a low frequency where the loop is liable to respond and pull out of the false lock. At higher and lower ratios, the hold-in range becomes wider and the beat note frequency becomes higher. Figure 8.17 shows waveforms for two ratios. When α is 2, the hold-in range covers the middle half of the characteristic (assuming the loop response to the beat note is negligible). When $\alpha = \frac{1}{2}$, as shown, or the reciprocal of any integer, the hold-in range for false lock is just as wide as for a true lock and

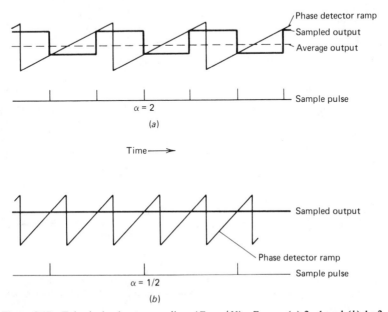

Figure 8.17 False locks due to sampling. $(F_{OUT}/N) : F_{REF} = (a)\ 2 : 1$ and $(b)\ 1 : 2$.

there is no beat note. Sampling every nth reference period produces a clean dc output. Without special circuitry to correct this problem, the false lock can only be absolutely prevented if it is not possible for the VCO to tune to the false-lock frequency.

8.6   SIMULATION OF ACQUISITION

While it is difficult to analyze exactly the acquisition of lock in a phase-locked synthesizer, including the effects of all of the important singularities and nonlinearities, the digital computer can do an exact simulation of acquisition. Such a simulation allows the designer to observe the response of the proposed design, and the effects of parameter variations, so he or she can predict performance and optimize design. One such program, called PH, has been written by the author and will be described here to illustrate the value of such a program.[21] The use of a commercially available program (CSMP III) will also be described.

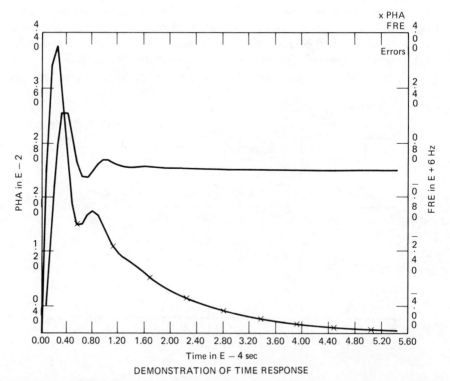

Figure 8.18 Graphic output of computer simulation for a loop with a sawtooth phase detector. FRE is output frequency error and PHA is phase error (i.e., offset from final value) response to a frequency step.

21   Egan, "Phase-locked loop simulation program."

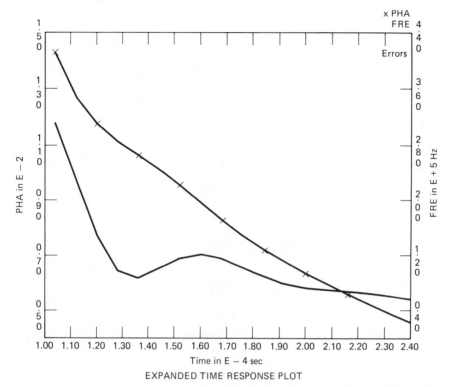

Figure 8.19 Expanded output of simulation. This display shows a small part of the previous display expanded to fill the screen. Reprinted with permission from *Microwaves*, Vol. 18, No. 5, copyright Hayden Publishing Co., Inc., 1979.

PH permits the user to enter a description of the loop and initial conditions and then to observe the loop response at sample instants. A practically arbitrary loop filter may be described by its poles and zeros and several types of phase detectors and associated acquisition aiding logic can be simulated. Phase-detector limitations, efficiency, maximum step amplitude, and cutoff of ramp amplifiers, may be specified.

The user accesses the program from a remote computer terminal. He or she is aided in the entry of data by a conversational format that will list parameter values when requested, permit them to be changed, and give their meaning when asked. Output may be obtained on a cathode ray tube (CRT), typewriter terminal, or the high-speed printer located at the computer. It may be in the form of data or graphs. Figures 8.18–8.21 show output originally displayed on a CRT and reproduced on an associated copier.* Previous data pages had recorded the parameters of the loop. As can be seen, the transient response can be plotted in several different ways.

The transient response of a phase-locked synthesizer can also be simulated

---

*Some features have been modified to make them more readable in print.

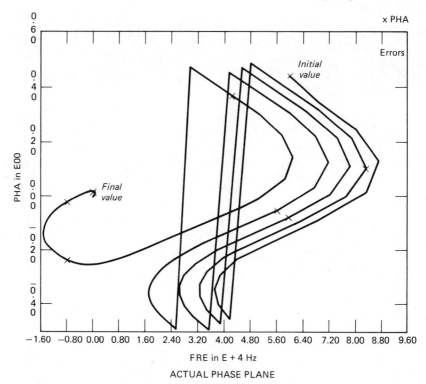

PHA in E00

FRE in E + 4 Hz

ACTUAL PHASE PLANE

Figure 8.20 Graphic output of computer simulation, PHAse (modulo 1) in cycles versus frequency in hertz. *Italics* added.

with the aid of a program such as CSMP III (Continuous System Model Program III, an IBM program product)[22, 23], which has the advantage of general availability. The output from a CSMP simulation is shown in Fig. 8.22 and the program is discussed in Appendix 8B. As with PH, true nonlinear response and signal-dependent sampling can be simulated with CSMP. CSMP can handle many nonlinearities whose effects would be quite difficult to predict otherwise.

8.7 TESTING FOR ACQUISITION

One way to determine acquisition parameters of a given configuration is to test the configuration, although this may only become practical later in the design process than we might desire.

Pull-in range can be determined by adding a bias which causes the operating point to shift far enough that the lock is broken. The bias can then

---

22 IBM.
23 Speckhart.

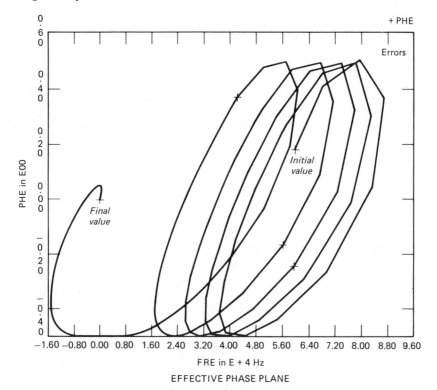

Figure 8.21 Graphic output of computer simulation, PHE (phase detector output) versus FRE (output frequency). *Italics* added.

be reduced and the point of lock observed; the mistuning at that point is one extreme of the pull-in range. A similar process would be used to find the other extreme. For example, a current can be introduced into a low-pass filter at the VCO input, as shown in Fig. 8.23, to offset the VCO tuning curve. This simulates a drift in VCO frequency, at constant tuning sensitivity. Alternately, a voltage might be added to the actual phase-detector output to give a constant tuning voltage, until it is out of hold-in range; then lock will be broken. As the added voltage is reduced, the phase at which lock is reacquired can be observed. These tests are equivalent as long as the elements between the points of bias introduction remain linear. A similar result may be obtained by changing the reference frequency if the change is small enough so the loop parameters are not significantly changed.

The time required for the frequency error to be reduced to a given value, after a frequency step, may be measured by observing the output of a frequency discriminator on an oscilloscope which is synchronized with the switching command. Alternately, the difference frequency between the synthesizer output and another signal at the synthesizer's final frequency may be observed. The period of the beat note may be measured or, if this is too great for the required resolution, the slope of the difference frequency signal,

PHASE LOCKED LOOP SIMULATION

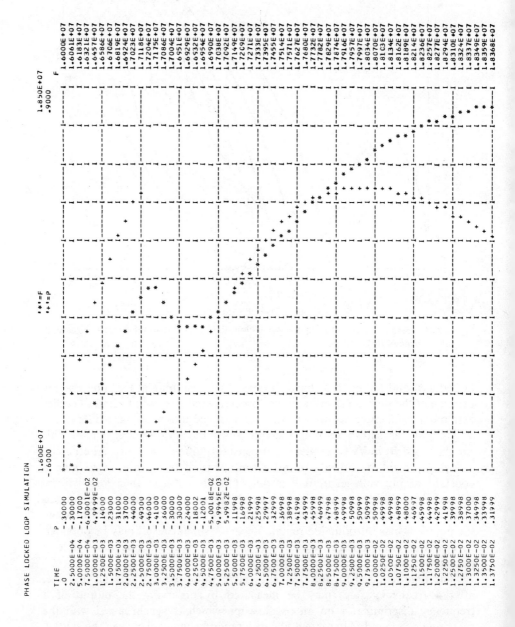

234

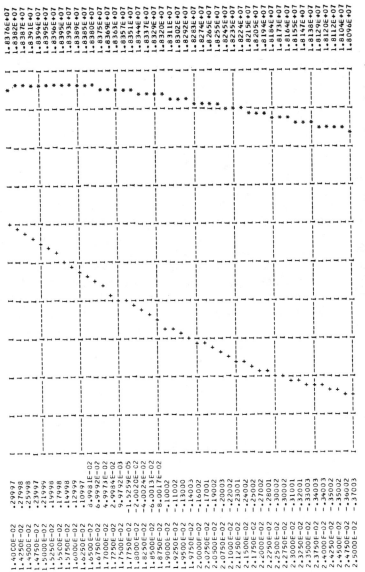

Figure 8.22 Graphical output from a CSMP simulation showing phase-detector output (+) and output frequency (∗) versus time.

235

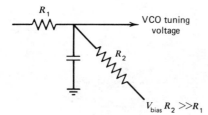

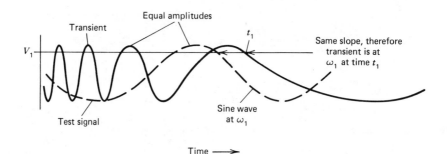

$V_{bias} R_2 \gg R_1$ **Figure 8.23**   Injection of a bias signal.

Figure 8.24   Measuring instantaneous frequency by comparison of slopes.

measured at a given voltage level, may be used to determine instantaneous frequency. This is illustrated in Fig. 8.24. Here a test signal has the same amplitude as the transient and its frequency is adjusted until both signals have the same slope at the same voltage level V_1. Since the amplitudes, voltages, and slopes of the test signal and the transient at t_1 are the same, their frequencies are also equal at that time.

APPENDIX 8A

Detailed Analysis of Simple Loop Acquiring Lock

See Section 8.1 for problem definitions.

   With a gain error, the equivalent of Eq. (8.10) for the nth period is

$$F_{OUT,n}(\varphi) = F_{ss} \exp(G_R \varphi_{nc}) = F_{ss} \exp\left[G_R\left(\varphi_{(n-1)c} + \Delta\varphi_{nc}\right)\right]. \quad (8.A.1)$$

Combining this with Eq. (8.17), we obtain

$$E_n = \frac{F_{ss}}{NF_{REF}} \exp\left[G_R\left(\varphi_{(n-1)c} + \Delta\varphi_{nc}\right)\right] - 1 \qquad (8.A.2)$$

$$= \frac{F_{OUT,(n-1)}}{NF_{REF}} \exp(G_R \Delta\varphi_{nc}) - 1. \qquad (8.A.3)$$

Using Eqs. (8.A.3) and Eq. (8.2), we obtain

$$E_n = \frac{F_{OUT,(n-1)}}{NF_{REF}} \exp\left[G_R\left(\frac{F_{REF}}{F_{s(n-1)}} - 1\right)\right] - 1. \qquad (8.A.4)$$

Using the expansion of the exponential for a small argument and retaining only the largest three terms in the expansion, the following is obtained:

$$E_n = \frac{F_{OUT,(n-1)}}{NF_{REF}}\left[1 + G_R\left(\frac{NF_{REF}}{F_{OUT,(n-1)}} - 1\right)\right.$$

$$\left. + \frac{G_R^2}{2}\left(\frac{NF_{REF}}{F_{OUT,(n-1)}} - 1\right)^2\right] - 1. \qquad (8.A.5)$$

Substituting $E_{(n-1)} + 1$ for $F_{OUT,(n-1)}/(NF_{REF})$, we obtain

$$E_n = -E_G E_{n-1} + \frac{1}{2}\frac{E_{(n-1)}^2}{1 + E_{(n-1)}} G_R^2. \qquad (8.A.6)$$

If

$$E_{n-1} \ll E_G, \qquad (8.A.7)$$

then only the first term in Eq. (8.A.6) need be retained. By iteration, we then obtain

$$E_n = -E_G E_{n-1} \qquad (8.A.8)$$

$$= -E_G(-E_G E_{n-2}) \qquad (8.A.9)$$

$$= -E_G\left[-E_G(-E_G E_{n-3})\right] \qquad (8.A.10)$$

$$= (-E_G)^n E_0. \qquad (8.A.11)$$

If, on the other hand, E_G is zero, Eq. (8.A.6) becomes

$$E_n = \frac{1}{2}\frac{E_{n-1}^2}{1 + E_{n-1}} \qquad (8.A.12)$$

$$\approx \tfrac{1}{2}E_{n-1}^2 \qquad (8.A.13)$$

for

$$E_{n-1} \ll 1. \qquad (8.A.14)$$

Iteration of Eq. (8.A.13) produces

$$\frac{E_n}{2} = \left(\frac{E_{n-1}}{2}\right)^2 \qquad (8.A.15)$$

$$= \left(\frac{E_{n-2}}{2}\right)^4 \qquad (8.A.16)$$

$$= \left(\frac{E_{n-3}}{2}\right)^8; \qquad (8.A.17)$$

$$E_n = 2\left(\frac{E_0}{2}\right)^{2n}. \qquad (8.A.18)$$

APPENDIX 8B

Transient Simulation with CSMP III

Figure 8.25 is a block diagram of the simulation which produced Fig. 8.22. The program is shown in Fig. 8.26.

Lines 20–115 of the program give parameter values and some simple preliminary calculations. Lines 140–240 describe the synthesizer loop.

Line 140 describes the reference ramp which begins at a specified initial phase and changes by one cycle each reference period. Each time it reaches 0.5 cycle, NI increments by one to return the reference phase to -0.5 cycle.

Line 150 describes the sample-and-hold function. Whenever the frequency divider output DI reaches one cycle, P becomes equal to the ramp value YR. DI is then recycled and the function ZHOLD, by definition, holds that value of YR until the divider output again reaches one cycle.

Lines 160–210 describe the serial connections of blocks which produce gain (KFG = K_F) and filtering and add constant offsets to give the desired initial frequency (F0) and final phase (PNL).

Line 220 represents the action of the frequency divider ($\div$ M) and line 230 integrates the divider output frequency to produce phase. The integrator output is given an initial value slightly greater than one in order to cause the hold function to acquire the initial phase.

Lines 260 and 265 increment the constants that recycle the frequency-divider output phase and the reference ramp, respectively, each time they change by one cycle. They are preceded by NOSORT and include the variable KEEP in order to insure that they are executed only once for each computation time, after the iterative solution for that instant has been completed.

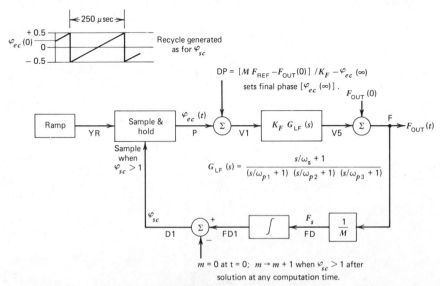

Figure 8.25 Block diagram of a CSMP simulation.

```
$$$CONTINUOUS SYSTEM MODELING PROGRAM  III   V1M3   TRANSLATOR OUTPUT$$$
                                                                       00000010
INITIAL                                                                00000020
PARAM ZERO=30.9, POLE1=8.09, POLE2=340., POLE3=400., KFG=.3E7, ...      00000030
    M=4500., PNL=-.3, FR=4.E3, PHO=-.3, FO=16.E6                        00000050
TP = 6.283185                                                          00000060
Z1=1/(TP*ZERO)                                                          00000070
P1 = 1/(TP*POLE1)                                                       00000080
P2 = 1/(TP*POLE2)                                                       00000090
P3 = 1/(TP*POLE3)                                                       00000100
DP = (M*FR-FO)/KFG - PNL                                                00000110
N= 0                                                                    00000115
N1 = 0                                                                  00000120
DYNAMIC                                                                 00000140
YR = FR * RAMP(0) + PHO - N1                                            00000150
P = ZHOLD(D1-1,YR)                                                      00000160
V1 = P + DP                                                             00000170
V2 = LEDLAG(Z1,P1,V1)                                                   00000180
V3 = KFG * V2                                                           00000190
V4 = REALPL(0,P2,V3)                                                    00000200
V5 = REALPL(0,P3,V4)                                                    00000210
F = FO + V5                                                             00000220
FD = F/M                                                                00000230
FD1 = INTGRL(1.00001,FD)                                                00000240
D1 = FD1 - N                                                            00000250
NOSORT                                                                  00000260
IF ((KEEP*D1).GT.1) N = N+1                                             00000265
IF ((KEEP*YR).GT.0.5) N1=N1+1                                           00000270
LABEL PHASE LOCKED LOOP SIMULATION                                     00000280
PRINT N,P,F,D1,YR                                                       00000290
OUTPUT P,F                                                              00000300
TIMER FINTIM = 25.E-3, DELT = 2.5E-6                                    00000310
METHOD RKSFX                                                            00000320
END                                                                     00000330
STOP
```

Figure 8.26 A CSMP program for simulating the transient response of a synthesizer loop.

Lines 270–290 specify the output; printed values are called for in addition to the graph shown in Fig. 8.22.

Line 300 specifies a 25-msec simulation with computations made each 2.5 μsec. Although only every 100th computed point (a default value) is displayed, increments between computation times must be small for accurate results. Unlike PH, which computes the duration of each period between samples, this CSMP simulation computes the state of the system each 2.5 μsec. Thus the sample might be late by this much or the reference phase at YR might be this late recycling. Also note that values are displayed at the steady-state sampling rate (other choices can be made) which here results in each sampled value of phase (P), except the first value, being displayed just once. These points really represent a series of levels of nearly equal length with the first two points occurring at opposite ends of the initial level.

The loop simulated in this example is relatively simple but there is no apparent reason why additional complexities, such as acquisition-aiding logic and limitations on phase-detector performance, cannot be added using CSMP III.

PROBLEMS

8.1 Refer to Fig. 8.2.

(a) Show that, if a small change in F_{REF} occurs between samples, F_{OUT}

will be at NF_{REF} two samples later if the loop has no filter and $\omega_0 = 1/T_s$. What happens if the change occurs at the sample time?

(b) Show that a small reference-phase step causes a frequency error for one period.

(c) Sketch the frequency versus time for (a) and (b).

8.2 A simple loop, such as in Fig. 8.2, operates at a 1-kHz reference frequency. Sketch the steady-state frequency at the divider output, giving values for the frequency, as a function of time, for the following loop gains:

(a) $\omega_0 = 1995$

(b) $\omega_0 = 2200$

(c) $\omega_0 = 3000$.

8.3 Give the best estimate available in the following, considering the applicability of the limits of various equations.

(a) What is the pull-in range for a loop with a sinusoidal phase detector and a lag-lead filter if the loop has the following parameters (assume no false locks):

$$N = 100,$$
$$K_F = 10^5,$$
$$\omega_z = 100,$$
$$\omega_p = 10?$$

(b) What is the seize frequency?

(c) What is the pull-in time from a 35-Hz initial frequency error?

(d) What is the seize time for a 10:1 frequency change?

8.4 For the loop of Problem 8.3, compute the pull-in frequency for a triangular and a sawtooth phase detector and compare to the results for a sinusoidal phase detector. Also, compare to the value obtained using Fig. 8.9.

8.5 A loop has a dc gain such that the total voltage range of a sawtooth phase detector ($\approx 360°$) would cause a frequency change of 10 kHz at the divider output. What is the pull-in range referenced to the divider output if the loop filter is a low-pass with the corner at

(a) 10 kHz;

(b) 1 kHz.

Assume no false locks. Use both the appropriate equations and the appropriate graph and compare the answers.

8.6 A loop has a perfect-integrator loop filter and the VCO output is processed through an IF filter. Assume the VCO frequency is centered in the filter and the filter consists of five identical isolated single-pole sections, each with a 400-kHz 3-dB bandwidth. How far will the nearest false-lock frequency be from the true lock frequency?

8.7 A loop is used to lock a 100-MHz VCXO to a 1-MHz reference. The loop bandwidth is to be 10 Hz. The phase detector produces 2 V linearly over 360°. The VCO output is disconnected from the divider and an external signal source is connected such that the phase detector skips cycles rapidly enough to produce essentially a dc voltage out of the loop filter. The dc offset is then adjusted to cause 100 MHz exactly out of the VCO. However, it is estimated that this voltage will vary ±0.1 V with temperature and aging.

   (a) If the loop filter has a pole at 0.1 Hz, where should the zero be for maximum pull-in range?

   (b) What is the maximum frequency offset (drift) for certain acquisition?

8.8 Derive Eq. (8.31A). Hint: see the footnote under the equation.

9
Acquisition Aids

In Chapter 8, we discovered various limitations that can prevent acquisition of lock. We also saw that it is difficult to predict exactly the range of parameters which will assure acquisition in many practical loops. In this chapter, we shall describe circuits to overcome these limitations and uncertainties.

There are four basic groups of such acquisition-aiding devices. The first type measures the frequency and uses this information to help tune the VCO to the correct frequency. The second type sweeps the VCO to help it acquire lock. The third type changes the loop parameters during acquisition. The fourth type employs logic to change the phase-detector output to something more appropriate for acquisition.

None of these circuits increases the hold-in range. They merely help to bring the VCO to the correct frequency so the phase-detector output can hold it there.

9.1 THE FREQUENCY DISCRIMINATOR AS AN ACQUISITION AID

The output of a frequency discriminator may be added to that of the phase detector to produce a correction voltage with the proper sense when the phase detector itself is skipping cycles and producing little correction signal.[1] Such an arrangement is shown in Fig. 9.1.

If, for some reason, the frequencies to be compared are not constant, a type of discriminator which produces a voltage proportional to the difference between two frequencies may be used. It is called a quadricorrelator.[2] A quadricorrelator might also be needed if the required accuracy of tuning were too great for a simple discriminator; for example, to help a 100 MHz VCO in

1 Gardner, p. 54.
2 Richman, "The DC quadricorrelator . . . ."

242

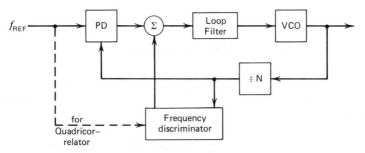

Figure 9.1 Acquisition aided by discriminator.

a loop with a 100 kHz pull-in range, a discriminator would have to maintain very accurate tuning.

The block diagram for a quadricorrelator is shown in Fig. 9.2. It mixes the two frequencies to be compared and differentiates the output at the difference frequency, thus establishing a proportionality between frequency difference and signal amplitude. The differentiated signal is synchronously detected in such a manner that the resulting "dc" has both amplitude and polarity representative of the frequency error. The differentiator may be a high-pass filter with transfer function.

$$G(\omega) = \frac{j\omega/\omega_1}{1 + j\omega/\omega_1} \approx j\frac{\omega}{\omega_1} \quad \text{for} \quad \omega \ll \omega_1, \tag{9.1}$$

where ω_1 is below the maximum difference frequency. An additional high-pass, with a corner near the phase-locked loop bandwidth, may be desirable, in series, to prevent the discriminator voltage from interfering with the normal pull-in process, once the difference frequency has become small enough.

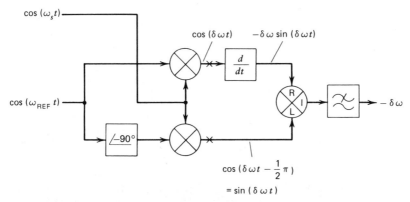

Figure 9.2 Quadricorrelator concept. Constant multipliers have been dropped. Filters may be added at points x to reduce noise and low-frequency response.

9.2   THE SEARCH OSCILLATOR AS AN ACQUISITION AID

A varying voltage can be added to the phase-detector output while that output is skipping cycles; the varying voltage can cause the VCO to tune past its steady-state frequency and thus come within the acquisition range of the unaided loop. If this added search voltage is changing slowly enough, the loop will lock and it can then maintain lock.

Such a system is shown in Fig. 9.3. The loop filter forms part of the search oscillator feedback circuit. Total feedback around amplifier A causes its output to oscillate and this, in turn, sweeps the VCO through its steady-state frequency. The oscillation may be sinsoidal, or positive feedback around A may be used to give a square wave at the output of A (making it a Schmidt trigger) which is converted to approximately a triangular sweep by a low-pass filter at G_S. When lock occurs, the oscillation must be stopped. This can be accomplished by the decreased loop gain caused by the entry of the phase-locked loop into the mathematical picture. The loop through A must be stable during phase lock and unstable otherwise. To the degree that the phase lock prevents the VCO tuning voltage from changing, it breaks the loop required for the sweeping oscillation. The effect of feedback through A, on the transfer function from point B to point C, must be included in the equivalent loop filter of the phase-locked loop, but, in the case of the Schmidt trigger, this feedback is effectively nullified.

Suppose the amplifier, A, is operated as a Schmidt trigger, changing states when the tuning voltage rises to V_U or falls to V_L. The tuning voltage is shown in Fig. 9.4 for the case where the phase-detector output is in the center of its range. If the shaded area reaches the required steady-state tuning voltage, the loop will lock. The phase detector will then add the necessary voltage to maintain the lock, as long as it is within the instantaneous hold-in range shown. The loop is capable of acquiring frequencies over the total range shown by the shaded area, which is indicated at A. The total hold-in range is

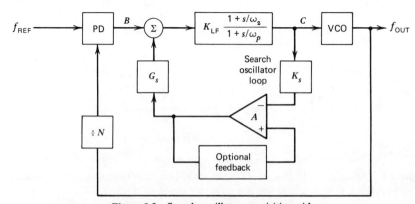

Figure 9.3   Search oscillator acquisition aid.

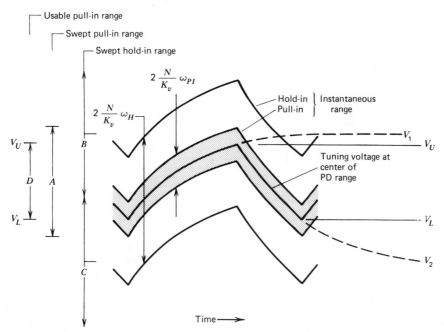

Figure 9.4 Swept tuning voltage when phase-detector characteristic is centered. Also shown are hold-in and pull-in ranges as the added voltage of the search oscillator sweeps.

the sum of B and C. They are centered on the two voltages V_1 and V_2, toward which the sweep relaxes. The usable pull-in range is restricted by V_U and V_L, as shown at D, since an attempt to reach beyond these limits will cause the sweep to reverse. If the instantaneous hold-in range were any smaller, or if V_1 and V_2 were further separated, there would be a hole in the middle of the hold-in range and lock could not be maintained at mid-range.

A change in the offset (imbalance) voltage appearing at the phase-detector output will cause V_1 and V_2 to change and may change the usable pull-in range. If such a change should bring V_1 or V_2 between V_U and V_L, the sweep would stop without the loop necessarily being locked.

The maximum sweep rate has been obtained for a type-one second-order loop with a sinusoidal phase detector when

$$\omega_0 \gg \omega_n \tag{9.2}$$

and

$$\zeta = 0.707. \text{ }^3 \tag{9.3}$$

For these conditions, probability of lock on one sweep is 100% if the sweep rate is

$$\dot\omega \leqslant \omega_n^2/2. \tag{9.4}$$

3 Gardner, p. 47.

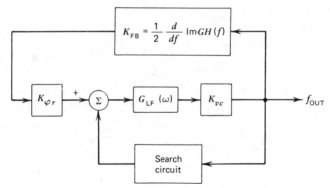

Figure 9.5 Feedback through the VCO during search.

Sweeping at twice this rate reduces lock probability to zero. Decreased damping is expected to decrease the allowed sweep rate.

As indicated by Eq. (8.66), skipping cycles can produce a voltage, out of the phase detector, which tends to false lock the loop. Just as the desired output from the phase detector stops the sweep oscillation, this voltage from a false lock can stop the sweep or prevent it from starting. A block diagram showing how this voltage is introduced is shown in Fig. 9.5. The search circuit design should be such that the search loop is stable with the feedback received during normal lock, but not with the feedback from possible false locks.

9.3 CHANGING LOOP PARAMETERS TO AID ACQUISITION

The loop can be converted to a simple loop without filter and the filter can be reinserted after acquisition. Figure 9.6 shows two circuits for doing approximately this, without disturbing the tuning voltage after acquisition.[4] In Fig. 9.6a, large signals, which occur when the loop is out of lock, pass through the back-to-back diodes and see a relatively small time constant while smaller signals, which occur in the locked state, see the full filter time constant. The nonlinear diode characteristic must be properly matched to the shunting resistance to produce a transition at the desired current level. In addition to the difficulty of doing this, there is the potential problem of large spikes from the phase detector causing the diodes to conduct in normal operation and thus defeat the filtering action. The circuit in Fig. 9.6b does not have these difficulties but does require that some signal change its state at the proper time.

4 Rey.

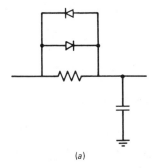

(a)

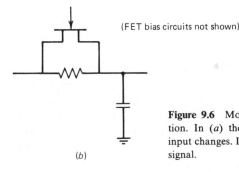

(FET bias circuits not shown)

(b)

Figure 9.6 Modifying loop-filter parameters to aid acquisition. In (a) the filter-corner frequency is higher for larger input changes. In (b) the bandwidth is switched by an external signal.

9.4 ACQUISITION-AIDING LOGIC

Probably the most useful devices for aiding acquisition in synthesizers are logic circuits which prevent the phase detector from producing the waveforms which typically occur during cycle skipping. Instead, a voltage is produced which causes acquisition to occur in an efficient manner. All of the methods to be described prevent false lock and the use of one of them is recommended whenever false lock is a potential problem.

9.4.1 Phase-Frequency Detector

We have already described the action of the phase-frequency detector as a phase detector (Section 5.1.6). Now we shall consider its acquisition-aiding features.

The phase detector has three possible output states, as shown in Fig. 9.7. The name of each state indicates the output which occurs in that state. At the occurrence of each reference pulse, the existing state is exited in the direction indicated by the arrow marked R and, at each divider output pulse, the existing state is exited along a path marked D. It is evident that, if the divider output frequency exceeds the reference frequency, the state must eventually alternate between zero and " $-$ " (unless the reference disappears, in which

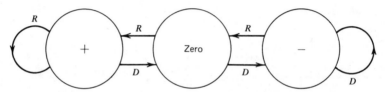

Figure 9.7 State transition diagram for phase-frequency detector.

case the occupied state will be steadily "−"). Conversely, a low divider output frequency results in the "+" state and zero state only ("+" only, if the divider stops).

Figure 9.8 shows typical outputs for these two cases and Fig. 5.18 illustrates, in another manner the characteristic that is traced out if the reference and divider output frequencies differ only slightly.

Consideration of Fig. 9.7 leads to the conclusion that, when the VCO is free running, this type of circuit will produce an average voltage of at least 50% of the maximum, that is, the equivalent of $\pm\pi$ rad. If the final phase is within this range, the frequency must be swept through its final value when the loop is out of lock and acquisition seems to be assured (although no proof is offered). When zero frequency error is reached, the phase will be between $-\pi$ and $+\pi$ rad. If the steady-state phase is zero (type-2 system), the frequency will overshoot to bring the phase to its final value.

It has been observed that, during cycle skipping, at least some practical realizations of this circuit generate pulses of the wrong polarity, when the inputs are nearly coincident. This limits the maximum useful frequency because the system will not function correctly if this erroneous output occurs over a significant part of the beat cycle.

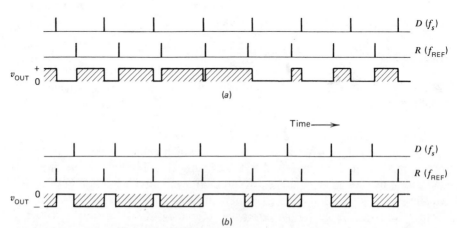

Figure 9.8 Output sequences for phase-frequency detector with frequency errors: (a) $f_{REF} > f_s$ and (b) $f_s > f_{REF}$.

Note that, during acquisition, the phase detector may be producing pulses much wider than those created at steady state and these, in addition to the transient response, could overload circuit components.

9.4.2 Divider Holding

A second type of acquisition-aiding logic has been incorporated in some integrated circuits.[5] We will call it "divider holding." The concept is illustrated in Fig. 9.9. The outputs from the reference and the variable dividers are forced to occur alternately. If the output from either divider is about to occur for a second time without an intervening output from the other divider, the divider which is about to produce the second consecutive output is held (stopped) until the other divider produces its output. The S-R flip-flop serves as a memory to indicate the last occurring output. At some count, shortly before the final count, an early count signal is generated. This is combined with the S-R flip-flop output in an AND gate such that a one is produced if the flip-flop indicates that the last final count was from the same divider that is now producing an early count. Thus, its final count is not produced until the other divider has produced a final count to change the state of the flip-flop. Since this results in the occurrence of two divider outputs in close sequence, it is apparent that the phase detector will operate near one extreme of its range. That extreme is at the correct end to bring the VCO on frequency.

Figure 9.10 illustrates the process by showing the counts of both the reference and preset dividers and the phase-detector output for several conditions. The phase-detector output is shown as a pulse whose duration is proportional to phase. However, both here and in the circuit to be described in the next section, the leading and trailing edges of this pulse could as well correspond to the start of a ramp and of a sample pulse, respectively (see Section 5.1.5).

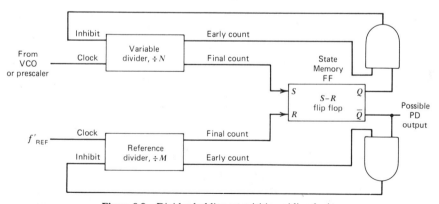

Figure 9.9 Divider-holding acquisition-aiding logic.

5 Fairchild . . . SH8096 . . . .

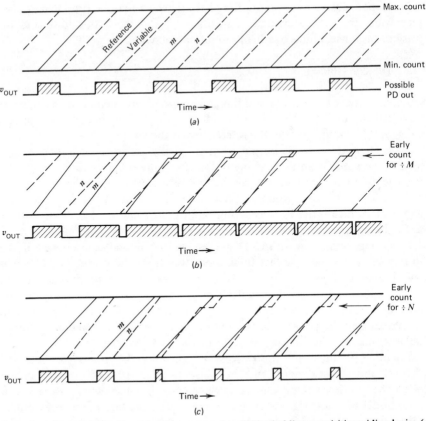

Figure 9.10 Counts and phase-detector outputs for divider-holding acquisition-aiding logic: (*a*) at phase lock, (*b*) with output frequency low, and (*c*) with output frequency high.

Figure 9.10*a* represents a phase-locked condition. The reference and variable dividers come to their terminal counts alternately and constant-width phase-detector pulses are produced. In Fig. 9.10*b*, the variable-divider output frequency is lower than the reference-divider output frequency. As the variable divider's final count comes later and later, the phase-detector pulse width widens. Finally, the early count of the reference divider occurs before the final count of the variable divider, causing the variable divider to be stopped for a time. This condition becomes repetitive under the conditions of constant frequency error for which the figure is drawn. In a closed loop, the high phase-detector output eventually causes the variable-divider output frequency to equal that of the reference-divider output. Then a linear-system pull-in begins, starting from zero frequency error and a well-defined (in contrast to the previous type) phase error. If the steady-state phase is close to the opposite extreme of the 360° phase-detector range, overshoot may be great enough to again bring into play the acquisition-aiding circuitry. This time it would be as shown in Fig. 9.10*c* for a high VCO frequency. However, assuming the loop is stable, it will produce less than 100% overshoot. Therefore, leaving one of the two out-of-lock conditions, the final lock value can be attained without entering the other. (This neglects complications such as changes in the sampling frequency and the fact that zero frequency error will not occur exactly coincident with the end of any count, but, ordinarily, these would not have much effect if reasonable margins are employed.)

One potential problem with this system is that, if the VCO should fail to oscillate due to a too-low tuning voltage, say at turn-on, the reference divider will be stopped at its early count to wait for an output from the variable divider. But, if the phase-detector output is low, the VCO may never start and, without a VCO signal, the variable divider will never reach its final count. This problem must be solved by preventing the VCO from stopping. However, even if it does not stop, there are some conditions under which a very low oscillator frequency can cause slowing of the response in this system.

9.4.3 Phase Range Limit[6]

This system is similar to the previous method except that the reference divider is never stopped. This is an advantage, when a common reference divider serves several loops, although not the only advantage, as will be shown.

The mechanization is represented in Fig. 9.11. When the preset divider frequency is high, it operates in the same manner as the "divider holding" system. But, when the preset-divider frequency is low, rather than stopping the reference divider to wait for the final count of the preset divider, a substitute final-count signal is generated. This changes the S-R flip-flop state and causes preset, just as would the actual final count. Thus, the sample period in the out-of-lock condition is always approximately equal to the reference period

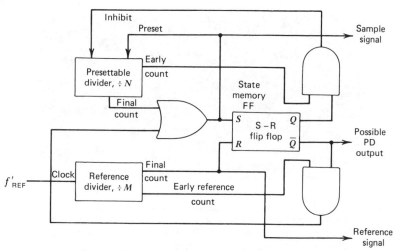

Figure 9.11 Phase range limit acquisition-aiding logic.

6 Egan, "Phase lock loop circuit."

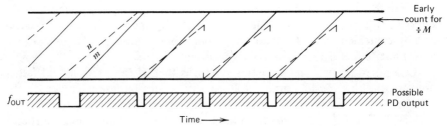

Figure 9.12 Counts and PD output for phase range limit acquisition-aiding logic with output frequency low.

and slowing of the response does not occur. Moreover, the correct voltage will be generated in the out-of-lock condition, even if the VCO stops.

Figure 9.12 shows the frequencies and the output of a flip-flop phase detector for the condition where the output frequency is low. Figure 9.13 shows actual outputs from a sample-and-hold phase detector during frequency steps.

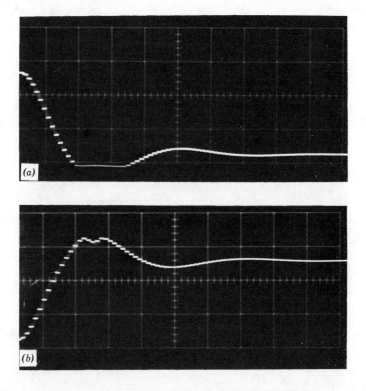

Figure 9.13 Frequency steps with phase range limit acquisition-aiding system observed on oscilloscope at (sample-and-hold type) phase-detector output: (*a*) negative step and (*b*) positive step. The dip in the peak may be associated with changes in phase as the frequency changes during the forced reset process and with changes from forced to normal reset and back. Forced reset can make the next sample pulse slightly earlier, resulting in a subsequent normal reset. This is followed by a forced reset, etc.

9.4.4   Reduced Linear Range for the Phase Detector

One of the principle advantages of the acquisition-aiding system described in Section 9.4.3 is that it permits a reduced linear range to be used in the phase detector, as illustrated in Fig. 9.14. With a fixed voltage range, often limited by supply voltage or component ratings, the phase-detector gain constant can be increased in inverse proportion to the portion of the reference period over which a linear characteristic is maintained, the ramp duty factor. The forward gain following the phase detector can then be proportionally reduced. This reduces the factor by which undesired noise components from the phase detector, both broadband noise and discrete multiples of the reference frequency, are multiplied, Thus, for example, if the ramp can be made steeper by a factor of four, by maintaining linear operation over only one quarter of the period, assuming the phase-detector noise remains unchanged, the resulting noise at the VCO goes down by a factor of four. To provide for efficient acquisition of lock with the reduced duty factor, the early-reference count from the reference divider is made to occur at the end of the linear range of the ramp, as shown in Fig. 9.14. Then, during acquisition when the VCO frequency is low, linear operation will begin at this phase.

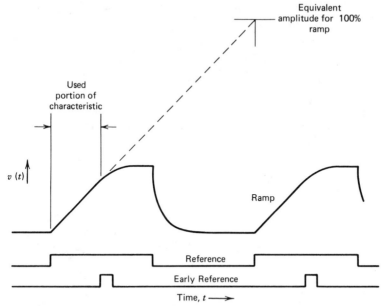

Figure 9.14   Phase-detector waveforms for operation with reduced duty factor.

PROBLEMS

9.1   Draw a mathematical control-system diagram corresponding to Fig. 9.1.

9.2   What is the maximum sweep rate for 100% probability of lock on one sweep, in Hz/sec, with the following parameters for a type-1, second-order loop?

$$f_p = 1 \text{ Hz,}$$

$$f_z = 71 \text{ Hz,}$$

$$f_0 = 10 \text{ kHz.}$$

9.3 Draw a diagram, similar to those in Fig. 9.8, for the phase-frequency detector, starting with the phase shown at the left of Fig. 9.8b, but for $f_s = \frac{3}{4}f_R$.

9.4 Draw a diagram showing phase detector-output, similar to those in Fig. 9.10, starting with the conditions shown at the right in Fig. 9.10b, but for a preset divider output frequency $\frac{1}{3}$ higher than the reference divider output frequency.

9.5 Draw a diagram similar to that of Fig. 9.12, and beginning identically to Fig. 9.12, but draw it for a system employing a 50% ramp duty cycle. In addition to the waveforms shown in Fig. 9.12, draw the ramp and the output from a sample-and-hold phase detector. The divider output frequency should be about $\frac{5}{6}$ of the reference frequency.

9.6 A synthesizer has -40-dB sidebands offset from the spectral center by the reference frequency. The phase-detector–ramp duty factor is reduced from 100% to 20% but the loop gain is kept constant. It is found that there is no change in the amplitude of the reference frequency signal which appears at the phase-detector output.

   (a) If the sidebands are caused by normal amplification of the phase-detector output (rather than by grounding, shielding, etc., problems), what will be their amplitude after the change?

   (b) If the sidebands are caused by a power-supply–isolation problem between the phase-detector output and the VCO, what will be the sideband amplitude after the change?

10

Spectral Purity

We have discussed the origin of noise sidebands within oscillators and how these are affected by the synthesizer and its components and also by passage through the system being driven by the synthesizer. In this chapter, we will discuss additional noise sources which must be controlled. We will also consider various ways in which spectral putiry is described.

The measure of spectral purity which we have used so far, spectral density, defines purity completely, but is not always the desired description. For example, many commercial synthesizers are specified according to the total phase noise within ± 15 kHz of spectral center, exluding the center 1 Hz. Or, there may be a requirement to know how much variation can be expected when the synthesized signal is counted repeatedly. These, and other, measures will be related to the spectral density so the synthesizer designer can interpret them in terms which will determine the required design.

10.1 NOISE SUPPRESSION[1-5]

It is difficult to overemphasize the importance of isolation of the control signal from noise sources in the phase-locked synthesizer. We have discussed, in Section 7.1, some of the precautions required to keep RF signals from causing undesired output components. We will not discuss that further at this point. Here, we reiterate the fact that a phase-locked synthesizer contains very sensitive circuits and very noisy circuits. The VCO, and the analog circuitry that drives its tuning voltage, are sensitive to disturbances from both within and without the synthesizer. The logic circuits, on the other hand, produce transient waveforms which can cause interference both within and without the synthesizer. The phase detector is caught in the middle. It must receive signals

---

1 Morrison.
2 White.
3 Manassewitsch, pp. 154–225.

4 Ott.
5 Motorola Data Sheet . . . MC4044.

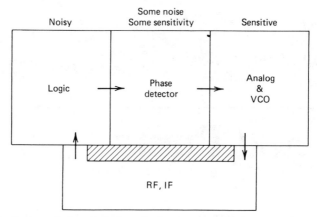

Figure 10.1   Regions of phase-locked synthesizer in terms of noise generation and sensitivity.

from the dirty world of the logic and deliver a signal to the clean world of the analog circuitry. The various regions of the systhesizer, in these terms, are illustrated in Fig. 10.1.

The following are some general suggestions to be judiciously applied in order to prevent noise contamination in phase-locked syntehsizers.

The regions shown in Fig. 10.1 should be separately shielded.

The logic should use a separate power supply from the sensitive circuits. Entry and exit from the logic should be through feedthrough filters designed to prevent escape of interference from the logic area. Care must be taken, however, to maintain required rise times on control signals.

Grounding and power bus routing in the phase detector should be analyzed and designed to prevent the mixing of logic signals with the phase-detector output.

All signal and ground wiring involving sensitive signals which pass through unprotected areas should be designed to minimize pickup.

Regulators or other active filters should be incorporated to prevent introduction of power-supply noise into sensitive circuits. All circuits, other than logic and RF and IF circuits, should be supplied from regulators. Modulation of RF and IF amplifiers by power supply noise is a possibility and regulation of these supplies might be required in some instances. Regulator requirements should be established based on expected line noise, circuit sensitivity, and spectral-purity requirements. It may be necessary to isolate the phase-detector power from the power for the other sensitive circuits to prevent transients, caused by generation of the ramp and by sampling, from modulating the synthesizer output. Care should be taken to insure regulator stability under all likely source and load conditions.

Power lines entering the phase detector and analog and VCO areas should be bypassed, preferably with feedthrough filters. RF and IF entries and exits can be protected by use of coaxial interconnections. Bypass capacitors from power supply lines to ground should be used to shorten the path of transient, RF and other signals.

Some noise sources cannot be eliminated by isolation. These are sources that are within the components of the loop and whose noise is transmitted along with the control signal. They include the noise sources discussed in Chapter 4 and jitter, discussed in Section 6.2. The signals in all types of synthesizers (indeed in all electronic systems) are subject to corruption by additive noise, including ever-present thermal noise, and must be kept sufficiently above the noise level throughout the synthesizers because the signal level cannot be improved relative to the noise level except by frequency division, filtering, or feedback.

10.2 SIDEBANDS CAUSED BY INJECTION LOCKING[6, 7]

When a signal is injected into an oscillator, the oscillator will tend to synchronize with that signal, that is, to operate at the same frequency. The smaller the frequency difference between the oscillator and the injected signal, and the greater the power of the injected signal, the more likely is synchronism. When the oscillator is synchronized by an injected signal, it is said to be injection locked. The VCO in a phase-locked synthesizer tends to be injection locked by coupled signals and their harmonics. High-speed logic signals are a particular threat, since they have many strong harmonics.

If the injected signal (including harmonics) does not have exactly the same frequency as the synthesized frequency, it can cause discrete sidebands to appear on both sides of the VCO output. Even if the frequency of the injected signal is the same as the required VCO frequency, FM sidebands can be caused by contention, for control of the phase, between the injected signal and the loop.[8] This is especially true in a type-two loop, which requires a zero phase-error signal.

10.3 CONVERSION OF PHASE NOISE DENSITY TO OTHER STABILITY MEASURES

A plot of phase power spectral density defines short-term stability in a way that permits calculations of other measures of short term stability. The process is not always reversable.

10.3.1 Phase and Frequency Deviation

What is the rms phase deviation for modulation frequencies between f_{m1} and f_{m2}? We consider the phase power spectral density distribution (see Section

6 Adler.
7 Kurokawa.
8 Underhill.

4.7) as representing phase deviations at many discrete modulation frequencies. The total mean-square deviation is the sum of the mean-square deviations at each frequency,

$$\tilde{\varphi}^2 = \sum_i \tilde{\theta}_i^2 \tag{10.1}$$

$$= \sum_i \delta_i S_\varphi(f_i), \tag{10.2}$$

where δ_i is one of the narrow frequency segments into which the band is divided. As the spacing and spectral width of the individual signals appraoch zero, this becomes

$$\tilde{\varphi}^2 = \int_{f_1}^{f_2} S_\varphi(f_m) \, df_m \tag{10.3}$$

and the rms phase deviation is the square root of this value.

A related question is: what is the rms frequency deviation? This is obtained similarly:

$$\tilde{\Delta f}^2 = \int_{f_i}^{f_2} S_{\dot{\varphi}c}(f_m) df_m \tag{10.4}$$

$$= \int_{f_1}^{f_2} f_m^2 S_{\varphi r}(f_m) \, df_m. \tag{10.5}$$

A similar question is: What is the sideband power in a bandwidth, Δf_1, excluding a band Δf_2, both centered on the spectrum? This is obtained from

$$P = 2 \int_{\Delta f_1/2}^{\Delta f_2/2} p(f_m) \, df_m, \tag{10.6}$$

where $p(f_m)$ is the average of the power density at $\pm f_m$ from center. Since curves of $S_\varphi, S_{\dot{\varphi}}$, or p can often be divided into straight-line segments on a log-log plot, we give the integrals in terms of the slopes of these segments. If the power spectral density from f_1 to f_2 has a slope of $b \times 10$ dB/decade ($b \times 3$ dB/octave), the power contained in that segment is

$$P = \frac{f_1 p(f_1)}{b+1} \left[\left(\frac{f_2}{f_1} \right)^{b+1} - 1 \right] \tag{10.7}$$

$$= \frac{f_2 p(f_2)}{b+1} \left[1 - \left(\frac{f_1}{f_2} \right)^{b+1} \right] \tag{10.8}$$

for

$$b \neq -1. \tag{10.9}$$

If, however,

$$b = -1, \tag{10.10}$$

Conversion of Phase Noise Density to Other Stability Measures

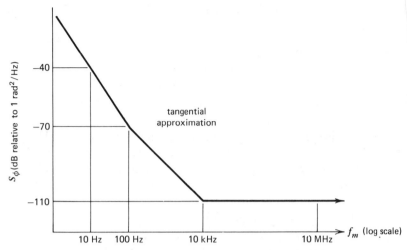

Figure 10.2 Phase spectral density for Example 10.1.

it is

$$P = f_1 p(f_1) \ln\left(\frac{f_2}{f_1}\right) \qquad (10.11)$$

$$= f_2 p(f_2) \ln\left(\frac{f_2}{f_1}\right). \qquad (10.12)$$

Of course, these equations apply equally to the integrals in Eq. (10.3) and (10.4) which involve S_φ and $S_{\dot\varphi}$, rather than p.

EXAMPLE 10.1

Problem For the phase power spectral density shown in Fig. 10.2, compute the rms phase deviation at modulation frequencies between 10 Hz and 10 MHz.

Solution For segment 1, $b = -3$. Using Eq. (10.7), we obtain

$$\varphi_1^2 = \frac{10 \text{ Hz} \times 10^{-4} \text{ rad}^2/\text{Hz}}{-3+1}\left[\left(\frac{100 \text{ Hz}}{10 \text{ Hz}}\right)^{-3+1} - 1\right]$$

$$= 4.95 \times 10^{-4} \text{ rad}^2.$$

For segment 2, $b = -2$. Using the same process as above, we obtain

$$\varphi_2^2 = \frac{100 \times 10^{-7}}{-1}(10^{-2} - 1) \text{ rad}^2$$

$$= 9.9 \times 10^{-6} \text{ rad}^2.$$

For segment 3, we have

$$\varphi_3^2 = 10^{-11}(10^7 - 10^4) \approx 10^{-4} \text{ rad}^2.$$

The sum of these three squared deviations is

$$\varphi^2 = 6.05 \times 10^{-4} \text{ rad}^2,$$

giving an rms deviation of

$$\varphi = 0.025 \text{ rad}.$$

10.3.2   Allan Variance

One popular measure of short-term stability is the mean-square value of the difference between adjacent (touching) frequency counts. When divided by $2f^2$, where f is average frequency, this is a special case of the Allan variance[9-11]. In symbols, the Allan variance for adjacent counts of length T is

$$\sigma_y^2(T) \triangleq \frac{\langle \Delta_i^2 \rangle_a}{2f^2}, \text{*} \tag{10.13}$$

where

$$\Delta_i \triangleq \bar{f}_{i+1} - \bar{f}_i, \tag{10.14}$$

$\langle \ \rangle$ indicates the average for all time, $\bar{f}_i$ is the average frequency over the nth time interval of duration T, and the subscript a means for adjacent counts. Generally, an estimate of $\sigma_y^2(T)$ is obtained by averaging a large number (M) of adjacent counts according to the equation

$$\sigma_y^2(T) \approx \frac{1}{2(M-1)f^2} \sum_{i=1}^{M-1} \Delta_i^2(T). \tag{10.15}$$

We will compute $\langle \Delta_i^2 \rangle$ in terms of S_φ for the more general case, where the time between the start of sequential counts is not necessarily equal to the length of the counts, and then specialize to the case of adjacent counts. The derivation is shown in full so it can serve as an example in relating other measures of spectral purity to density.

---

9   Allan.
10   Howe, p. 15.
11   Shoaf, pp. 50 and 51.
*$\sigma_y^2(T)$ is the normalized (to f) expected value of the sample variance for two samples, that is

$$\sigma_y^2(T) = \frac{1}{f^2} \left\langle \left[\frac{1}{M-1} \sum_{i=1}^{M} \left[\bar{f}_i - \frac{1}{M} \sum_{i=1}^{M} \bar{f}_i \right] \right]^2 \right\rangle_{M=2}$$

$$= \frac{1}{f^2} \left\langle \left[\bar{f}_2 - \tfrac{1}{2}(\bar{f}_1 + \bar{f}_2) \right]^2 + \left[\bar{f}_1 - \tfrac{1}{2}(\bar{f}_1 + \bar{f}_2) \right]^2 \right\rangle$$

$$= \frac{1}{f^2} \left\langle \tfrac{1}{4}(\bar{f}_1 - \bar{f}_2)^2 + \tfrac{1}{4}(\bar{f}_2 - \bar{f}_1)^2 \right\rangle$$

$$= \frac{1}{2f^2} \left\langle (\bar{f}_1 - \bar{f}_2)^2 \right\rangle.$$

Let the instantaneous frequency be

$$f = f_o + \sqrt{2} \sum_n \tilde{\delta}_n \cos(2\pi f_n t + \alpha_n), \tag{10.16}$$

where f_n is the modulation frequency of a noise component with rms deviation $\tilde{\delta}_n$ and the noise component represents the frequency deviation in a band sufficiently narrow that the amplitude is essentially constant for times of importance in the problem. Note that zero frequency drift has been assumed.

The average frequency in the first measurement period, of duration T_A, is

$$\bar{f}_1 = \frac{1}{T_A} \int_0^{T_A} f\,dt \tag{10.17}$$

$$= f_o + \frac{1}{\sqrt{2}\,\pi T_A} \sum_n \frac{\tilde{\delta}_n}{f_n} \sin(2\pi f_n t + \alpha_n) \Big|_0^{T_A} \tag{10.18}$$

$$= f_o + \frac{1}{\sqrt{2}\,\pi T_A} \sum_n \{ \tilde{\varphi}_{nr}[\sin(2\pi f_n T_A + \alpha_n) - \sin\alpha_n] \}, \tag{10.19}$$

where $\tilde{\varphi}_{nr}$ is the rms phase deviation in the above-defined bandwidth, expressed in radians.

For the second measurement period, which begins T_S after the start of the first period, the average is

$$\bar{f}_2 = \frac{1}{T_A} \int_{T_S}^{T_A + T_S} f\,dt \tag{10.20}$$

$$= f_o + \frac{1}{\sqrt{2}\,\pi T_A} \sum_n (\tilde{\varphi}_{nr}\{\sin[2\pi f_n(T_A + T_S)+\alpha_n]$$

$$- \sin(2\pi f_n T_S + \alpha_n)\}). \tag{10.21}$$

We then subtract these two averages and employ several trigonometric identities, as follows:

$$\Delta_1 = \bar{f}_2 - \bar{f}_1 \tag{10.22}$$

$$= \frac{1}{\sqrt{2}\,\pi T_A} \sum_n \tilde{\varphi}_{nr}\{\sin[2\pi f_n(T_A + T_S)+\alpha_n]$$

$$- \sin(2\pi f_n T_S + \alpha_n) - \sin(2\pi f_n T_A + \alpha_n) + \sin\alpha_n\} \tag{10.23}$$

$$= \frac{\sqrt{2}}{\pi T_A} \sum_n \tilde{\varphi}_{nr}\{\cos[2\pi f_n(\tfrac{1}{2}T_A + T_S) + \alpha_n]\sin(\pi f_n T_A)$$

$$- \cos(\pi f_n T_A + \alpha_n)\sin(\pi f_n T_A)\} \tag{10.24}$$

$$= -\frac{2\sqrt{2}}{\pi T_A} \sum_n \tilde{\varphi}_{nr} \sin(\pi f_n T_A)\sin(\pi f_n T_S)\sin[\pi f_n(T_A + T_S)+\alpha_n]. \tag{10.25}$$

The average of Δ_i^2 for all time is the same as the average of Δ_i^2 for all values of α_n from 0 to 2π:

$$\langle \Delta_i^2 \rangle = \frac{1}{2\pi} \int_0^{2\pi} \Delta_i^2 \, d\alpha \tag{10.26}$$

$$= \frac{4}{\pi^3 T_A^2} \sum_n \sin^2(\pi f_n T_A) \sin^2(\pi f_n T_S)$$

$$\times \tilde{\varphi}_{nr}^2 \int_0^{2\pi} \sin^2\left[\pi f_n (T_A + T_S) + \alpha_n\right] d\alpha_n \tag{10.27}$$

$$= \frac{4}{\pi^2 T_A^2} \sum_n \tilde{\varphi}_{nr}^2 \sin^2(\pi f_n T_A) \sin^2(\pi f_n T_S). \tag{10.28}$$

Reducing the band in which $\tilde{\varphi}_{nr}$ is measured to differential width, $\tilde{\varphi}_{nr}^2$ now becomes

$$\tilde{\varphi}_{nr}^2 \to S_{\varphi r}(f_n) \, df_n, \tag{10.29}$$

and the variance of Δ_i is

$$\langle \Delta_i^2 \rangle = \left(\frac{2}{\pi T_A}\right)^2 \int_0^{f_{max}} S_{\varphi r}(f_n) \sin^2(\pi f_n T_A) \sin^2(\pi f_n T_S) \, df_n. \tag{10.30}$$

If the counts are in adjacent periods, then

$$T \triangleq T_A = T_S \tag{10.31}$$

and

$$\langle \Delta_i^2 \rangle_a = \left(\frac{2}{\pi T}\right)^2 \int_0^{f_{max}} S_{\varphi r}(f_n) \sin^4(\pi f_n T) \, df_n. \tag{10.32}$$

To obtain $\langle \Delta_i^2 \rangle$ from a density plot, one plots the product of $S_{\varphi r}$ (in rad^2/Hz) and the appropriate sine function(s), integrates under the resulting curve, and multiplies the result of the integration by $[2/(\pi T_A)]^2$. An appropriate calculator program can be most helpful for performing the point to point integration.

Note that, after the first few nulls in the $\sin^4$ function, it will usually be changing much faster than S_φ. When this is true, the average value of $\sin^4$ may be used in the integration:

$$\int_{f_{HI}}^{f_{max}} S_\varphi \sin^4(\pi f_n T) \, df_n \approx \left(\frac{1}{\pi} \int_0^\pi \sin^4\theta \, d\theta\right) \int_{f_{HI}}^{f_{max}} S_\varphi \, df_n \tag{10.33}$$

$$= \frac{3}{8} \int_{f_{HI}}^{f_{max}} S_\varphi \, df_n \tag{10.34}$$

for

$$f_{max} - f_{HI} \gg \frac{1}{T} \tag{10.35}$$

and

$$f_{HI} \gg \frac{1}{T}. \tag{10.36}$$

Note also that certain noise frequencies are at nulls of the sine function and will not be seen in this measure of purity.

If discrete components are present at frequenciss $f_1, \ldots, f_i, \ldots, f_m$, then the variance may be written

$$\langle \Delta_i^2 \rangle_a = \left(\frac{2}{\pi T} \right)^2 \left[\int_0^{f_{max}} S'_{\varphi r}(f_n) \sin^4(\pi f_n T) \, df_n + \sum_{k=1}^{M} \tilde{\varphi}_{kr}^2 \sin^4(\pi f_k T) \right], \quad (10.37)$$

where $\tilde{\varphi}_{kr}$ is the rms deviation of the kth discrete component and $S'_{\varphi r}$ equals $S_{\varphi r}$ less the discrete components.

Tables are available[12] which give factors that can be used to convert between $\sigma_y^2(T)$ and $S_{\varphi r}$ when the slope of $S_{\varphi r}$, on a log-log plot, is constant.

10.3.3 Calculations of Other Time-Domain Stability Measures

The same method employed in the previous paragraph can be used to derive other relationships between time-domain stability measures and phase power spectral density. Some examples of these relationships follow. Again, no drift component is assumed.

What is the variance σ_b^2 of the frequency change over a period T?

$$\sigma_b^2 \triangleq \langle [\,f(t+T) - f(t)\,]^2 \rangle = 4 \int_0^{f_{max}} S_{\dot{\varphi}} \sin^2(\pi f_n T) \, df_n. \quad (10.38)$$

What is the variance σ_d^2 of the frequency during a period T?

$$\sigma_d^2 \triangleq \langle [\,f(t) - \bar{f}\,]^2 \rangle \Big|_t^{t+T} = \int_0^{f_{max}} S_{\dot{\varphi}} \left[1 - \left(\frac{\sin \pi f_n T}{\pi f_n T} \right)^2 \right] df_n. \quad (10.39)$$

10.4 MEASUREMENT OF FREQUENCY STABILITY

The Allan variance can be measured with a computing counter (e.g., HP5360A) which is programmed to make sequential counts and compute the resulting variance according to Eq. (10.15). Some small time is required between counts.

Sideband power density can be most conveniently measured by a spectrum analyzer if its sensitivity, dynamic range,* and spectral purity are great enough. Often they are not, and other methods, which we are about to describe, are needed.[13, 14]

Care should be exercised to prevent the test from changing the performance of the source being tested. The load seen by the oscillator should be similar to that to be seen in practice. Sufficient isolation should be maintained

---

12 Howe, p. 18.

*The ability to accurately measure weak signals in the presence of strong signals.

13 Reynolds.

14 Owen.

to prevent test signals from influencing the oscillator, especially by injection locking.

10.4.1 Measurement With a Phase Detector

Phase deviation can be measured by comparing two signals at identical average frequencies in a phase detector. A balanced mixer can be used for a phase detector and a loop may be used to lock one oscillator to the other to maintain the proper phase relationship for phase detection. Well above the loop bandwidth, the measured phase power spectral density is the sum of the density from the two oscillators. Below the loop bandwidth, the phase deviation is attenuated by the loop. The resulting phase-detector output may be measured as a function of frequency by a selective voltmeter or wave analyzer. The setup is diagrammed in Fig. 10.3.

By Eq. (5.1), the low-frequency output from the phase detector is approximately

$$v = A \cos\delta_{\varphi r}, \tag{10.40}$$

where

$$\delta_\varphi = \varphi_1 - \varphi_2.$$

By allowing a small difference frequency, this characteristic can be observed on an oscilloscope as δ_φ changes. This permits calibration of the phase detector. Depending on the value of δ_φ, the balanced mixer acts either as an amplitude or a phase detector.

$$\frac{dv}{dA} = \cos\delta_{\varphi r}, \tag{10.41}$$

$$dv = dA \text{ at } \delta_{\varphi r} = 0; \tag{10.42}$$

$$\frac{dv}{d\delta_\varphi} = -A \sin\delta_{\varphi r}, \tag{10.43}$$

$$dv = A \, d\delta_{\varphi r} \text{ at } \delta_{\varphi r} = -\pi/2. \tag{10.44}$$

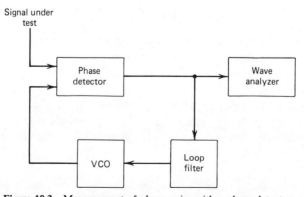

Figure 10.3 Measurement of phase noise with a phase detector.

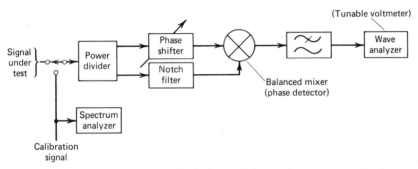

Figure 10.4 Basic block diagram of a single-signal phase-noise measurement system.

At normal mixer operating levels, A will be proportional to the amplitude of the weaker input signal and, therefore, at $\delta_\varphi = 0$, the balanced mixer detects AM on the weaker signal. At phase quadrature, the balanced mixer's output is proportional to phase.

Under certain conditions, a balanced mixer can be used to measure AM or FM with only one signal. If the LO is a strong, limited version of the other input signal, AM can be linearly detected when the two inputs are in phase. FM can be detected with only one signal if the FM sidebands are accentuated on one of the two quadrature inputs, both of which have been derived from the common signal. This has been done by notching out the carrier, with a high-Q cavity[15], at one of the inputs. The test setup is shown in Fig. 10.4. It can be calibrated by introducing a signal with small, discrete FM sidebands whose magnitude can be measured by a spectrum analyzer.

10.4.2 Measurement With a Frequency Discriminator

A frequency discriminator can be used to measure frequency deviation, as in Fig. 10.5. Since phase deviation falls 6 dB/octave faster than frequency deviation, this method tends to be more sensitive at higher modulation

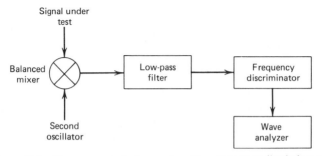

Figure 10.5 Measurement of phase noise with a frequency discriminator.

15 Ondria.

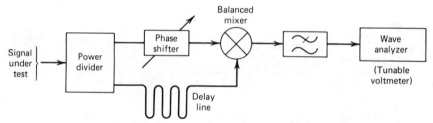

Figure 10.6 Basic block diagram of a discriminator using a delay line.

frequencies. Another signal may be required to convert the signal to be measured to a frequency where a discriminator is available.

One form of discriminator that is useful at microwave frequencies[16] is shown in Fig. 10.6. Here, the signal to be measured is divided, and one branch goes directly to the phase detector while the other branch is delayed before reaching the phase detector. The phase-detector output is

$$v = A \cos(\theta_1 - \theta_2). \tag{10.45}$$

If the additional delay T in path 2 is short compared to the modulation periods of interest ($\omega_m T \ll 1$), then the frequency can be considered constant throughout the delay line, giving

$$\theta_1 = \varphi + \omega(t + T) \tag{10.46}$$

and

$$\theta_2 = \varphi + \omega t. \tag{10.47}$$

Thus the output is

$$v = A \cos \omega T, \tag{10.48}$$

and, if the frequency is written as an average value, $\bar{\omega}$, plus a deviation, δ_ω, the the voltage is

$$v = A \cos(\bar{\omega} + \delta_\omega)T \tag{10.49}$$

$$= A(\cos \bar{\omega}T \cos \delta_\omega T - \sin \bar{\omega}T \sin \delta_\omega T). \tag{10.50}$$

When T is adjusted, using a line stretcher or phase shifter, so $\bar{\omega}T$ is an odd multiple of $\pi/2$, the first term disappears and, for small deviation ($T\delta_\omega \ll \pi/2$), the remaining term gives

$$v = \pm A \sin \delta_\omega T \approx \pm A T \delta_\omega. \tag{10.51}$$

While this shows that sensitivity is proportional to T, A also decreases with T, and amplifiers (which must introduce little noise, especially noise modulation) may be required to maintain the signal level at the phase detector as the delay is increased. Although the sensitivity is also proportional to A, the maximum

16 Lance.

usable value of A is limited by the phase detector, normally a balanced mixer.

The response at higher modulation frequencies ($\omega_m T \gtrsim 1$) is multiplied by $(\sin \frac{1}{2}\omega_m T)/(\frac{1}{2}\omega_m T)$.[17]

10.4.3   Measurement of Sideband Density

A method suitable for measuring sideband power density is shown in Fig. 10.7. This is basically a high-dynamic-range receiver that is tuned to the center of the spectrum, with appropriate attenuation inserted to prevent saturation, to measure total signal strength. It is then tuned off by f_m to measure signal strength at f_m from center. A good filter is required before the first amplifier to prevent its saturation, by the main spectral power, when the source has been detuned and attenuation has been removed. Again, the single-sideband power of the two mixed signals is measured. If the signal being measured is too unstable to be contained within the filter bandwidths, it may be necessary to measure its power with a power meter. The setup can then be calibrated against the power meter by substituting a stable source for the unstable source and comparing readings at the power meter and the wave analyzer.

If the main power is measured on the wave analyser as P_M, with an attenuation factor of $A_M \geq 1$, and the sideband power is measured as P_S, with an attenuation factor of $A_S \geq 1$, then the relative single-sideband power, in the measurement bandwidth, Δf, is

$$\mathcal{L}(f_m)\Delta f = \frac{P_S A_S}{P_M A_M} . \tag{10.52}$$

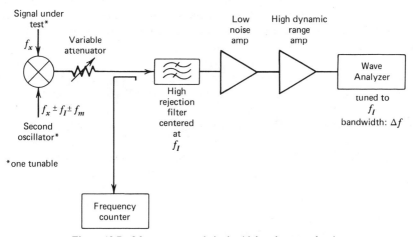

Figure 10.7   Measurement of single-sideband power density.

17 Manassewitsch, p. 125.

PROBLEMS

10.1 Repeat Example 10.1 but use power spectral densities that are 10 dB lower and corner frequencies that are 10 times higher. In other words, in Fig. 10.2, decreases each value on the ordinate axis by 10 dB and increase each value on the absicissa axis by 10 times. Start at 100 Hz.

10.2 Find the Allan variance σ_y (10 msec) due to the following discrete FM sidebands on a 100-MHz signal:
  -30 dBc at ± 900 Hz (dBc is relative to the carrier),
  -40 dBc at ± 1050 Hz.

10.3 Find the Allan variance σ_y (1 msec) due to phase-noise spectrum, on a 1-GHz signal, which has a density of 10^{-7} rad^2/Hz from 0 to 50 kHz and then (the phase power spectral density) falls at 20 dB/decade. This problem can be done without any detailed graphical construction.

10.4 Two crystal oscillators are compared in a phase detector. Oscillator L is limited in amplitude and then applied to the LO port of the doubly balanced mixer. The other, lower-level oscillator, R, is applied to the RF port. The IF output of the mixer is low passed to remove the sum and input frequencies and the difference frequency is displayed on an oscilloscope. A sine wave with 10-mV peak-to-peak amplitude appears. Furthermore, the average value of the sine wave is zero. One oscillator is then phase locked to the other with a 10-Hz loop bandwidth and the output is put into a wave analyzer (tunable voltmeter) with a 10-Hz bandwidth and tuned to 1 kHz. When the phase of one input signal is adjusted for maximum DC level out of the phase detector, the wave analyzer measures 150 μV rms. When the phase is adjusted for a null at the output, the analyzer measures 60 μV rms. From this, what is known about the spectral characteristics of the two oscillators?

10.5 In Fig. 10.7, the filter is centered at 10.7 MHz and is 10 kHz wide. The attenuator is set at 50 dB and the wave analyzer reads 0 dBm, in either a 1- or 5-kHz bandwidth (at the analyzer), when the counter reads 10.7 MHz. When the counter reads 10.9 MHz, and the attenuator is set to 0 dB, the analyzer reads -75 dBm in a 200-Hz analyzer bandwidth. Assuming that one of the input signals has much better spectral purity than the other, what is known from the above about the second (less clean) oscillator?

Bibliography

Abe, Hiroyuki *et al.* "A Highly Stabilized Low-Noise GaAs FET Integrated Oscillator with a Dielectric Resonator in the *C* Band." *IEEE Transactions on Microwave Theory and Techniques*, Vol. MTT-26, No. 3 (March 1978), pp. 156–162.

Adler, Robert. "A Study of Locking Phenomena in Oscillators." *Proceedings of IEEE*, Vol. 61, No. 10 (October 1973), pp. 1380–1385.

Allan, D. W. "Statistics of Atomic Frequency Standards." *Proceedings of IEEE*, Vol. 54, No. 2 (February 1966), pp. 221–230.

Alonzo, G. "Considerations in the Design of Sampling-Based Phase-Lock-Loops." *WESCON/66 Technical Papers, Session 23,* Western Electronic Show and Convention, 1966, part 23/2.

"A 1 GHz Prescaler Using GPD Series Thin-Film Amplifier Modules." *Microwave Component Applications, ATP-1036.* Avantek, Inc., 3175 Bowers Avenue, Santa Clara, CA 95051, 1977.

Beach, Richard M. "Hyperabrupt Varactor Tuned Oscillators." *Watkins-Johnson Co., Tech-Notes*, Vol. 5, No. 4 (July/August 1976). Also as "High Speed Linear Oscillators." *Microwave Journal* (December 1978) pp. 59–65.

Bearse, S. "TED Triode Performs Frequency Division." *Microwaves* (November 1975), p. 9.

Blanchard, Alain. *Phase-Locked Loops*. New York, NY: John Wiley, 1976.

Blood, W. R. Jr. *MECL System Designer's Handbook*. 2d ed. Mesa, AZ: Motorola Semiconductor Products, Inc., 1972.

Bracewell, Ron. *The Fourier Transform and Its Applications*. New York, NY: McGraw-Hill, 1965.

Braymer, N. B. "Frequency Synthesizer." United States Patent 3,555,446, January 12, 1971.

Breeze, Eric G. "High Frequency Digital PLL Synthesizer." *Fairchild Journal of Semiconductor Progress* (November, December 1977). Mountain View, CA: Fairchild Semiconductor, pp. 11–14.

Buchanan, J. "Dielectric Absorption—It Can Be a Real Problem in Timing Circuits." *EDN* (January 20, 1977), pp. 83–86.

Buswell, R. "Linear VCO's." *Watkins-Johnson, Co., Tech-Notes*, Vol. 3, No. 2 (March /April 1976).

Buswell, R. "Voltage Controlled Oscillators in Modern ECM Systems." *Watkins-Johnson, Co., Tech-Notes*, Vol. 1, No. 6 (December 1974).

Byers, W. *et al.* "A 500 MHz Low-Noise General Purpose Frequency Synthesizer." *Proceeding of the Twentieth Annual Frequency Control Symposium*. U.S. Army Electronic Command, Fort Monmouth, NJ, 1973.

Byrne, C. J. "Properties and Design of the Phase Controlled Oscillator With a Sawtooth Comparator." *The Bell System Technical Journal* (March, 1962), pp. 559–602.

Cahn, C. R. "Piecewise Linear Analysis of Phase-Locked Loops." *IRE Transactions on Space Electronics and Telemetry*, SET-8, No. 1 (March 1962), pp. 8–13. (Also in Lindsey, pp. 8–13.)

Clark, R. J., and D. B. Swartz. "Take a Fresh Look at YIG-Tuned Sources." *Microwaves* (February, 1972).

Cooper, H. W. "Why Complicate Frequency Synthesis?" *Electronic Design* (July 19, 1974), pp. 80–84.

Dana Series 7000 Digiphase® Frequency Synthesizers, Publication 980428 (Manual). Dana Laboratories, Inc., 2401 Campus Drive, Irvine, CA 92664, 1973.

Develet, J. A. Jr. "The Influence of Time Delay on Second-Order Phase-Lock Loop Acquisition Range." *Proceedings International Telemetering Conference*, Vol. 1 (Sept. 23–27, 1963), pp. 432–437. (From Lindsey, pp. 126–131.) Note: ϕ_0 should be θ_0 above Eq. 10.

Driscoll, M. "Two-Stage Self-Limiting Series Mode Type Quartz-Crystal Oscillator Exhibiting Improved Short-Term Frequency Stability." *IEEE Transactions of Instrumentation and Measurement*, IM-22, No. 2 (June 1973), pp. 130–138.

Egan, W. F. "LOs Share Circuitry to Synthesize 4 Frequencies." *Microwaves* (May 1979), pp. 52–65.

Egan, W. (ed.). *Miniature Low Noise Frequency Synthesizer* . . . , GTE Sylvania Report G1004 (June 1972). Internal publication.

Egan, W. F. "Phase-Locked Loop Simulation Program." *Proceedings of the 1976 GTE Symposium on Computer Aided Design*, Vol. 1. GTE Laboratories, Waltham, MA. June 1976, pp. 239–253. Internal publication.

Egan, W. F. "Phase Lock Loop Circuit." United States Patent 4,001,713, January 4, 1977.

Egan, W. and E. Clark. "Test Your Charge-Pump Phase Detectors." *Electronic Design*, Vol. 26, No. 12 (June 7, 1978), pp. 134–137. See also comment from J. Hatchett in *Electronic Design*, Vol. 26, No. 22 (Oct. 25, 1978), p. 14.

Fairchild Data Sheet: "Phase/Frequency Detector, 11C44", 6 p., Fairchild Semiconductor, Mountain View, CA.

Fairchild Preliminary Data Sheet: "SH8096 Programmable Divider—Fairchild Integrated Microsystems" (April 1970).

Fogarty, J. D. "Digital Synthesizers, . . . " *Computer Design* (July 1975), pp. 100–102.

Funk, R. "Low-Power Digital Frequency Synthesizers Utilizing COS/MOS IC's," Application Note ICAN-6716. RCA Solid State Division, Sommerville, NJ, 08876, March 1973.

Gardner, F. *Phaselock Techniques*. New York, NY: John Wiley, 1966.

Gibbs, J. and R. Temple. "Frequency Domain Yields Its Data to Phase-Locked Synthesizer." *Electronics (April 27, 1978), pp. 107–111.*

Gillette, G. *"The Digiphase Synthesizer." Frequency Technology* (August 1969), pp. 25–29.

Goldman, S. *Frequency Analysis, Modulation and Noise*. New York, NY: McGraw-Hill, 1948, pp. 172–175.

Goldstein, A. Jay. "Analysis of the Phase-Controlled Loop With a Sawtooth Comparator." *The Bell System Technical Journal* (March 1962), pp. 603–633.

Goldwasser, W. J. "Design Shortcuts for Microwave Frequency Dividers." *The Electronic Engineer* (May 1970), pp. 61–65.

Goodman, A. "Increasing the Band Range of a Voltage-Controlled Oscillator." *Electronic Design* (September 28, 1964), pp. 28–35.

Gorski-Popiel, J. (ed.). *Frequency Synthesis: Techniques and Applications*. New York, NY: IEEE Press, 1975. Includes contributions by Hutchinson (pp. 25–45) and Tierney (pp. 121–149) listed below.

Greenstein, L. J. "Phase-Locked Loop Pull-In Frequency." *IEEE Transactions on Communications*, Vol. COM-22 (Aug. 1974), pp. 1005–1013. (From Lindsey, pp. 150–158.) Note: Eq. 20 does not match its source precisely.

Grove, Wayne M. "A D.C. to 12 GHz Feedthrough Sampler for Oscilloscopes and Other R. F. Systems." *Hewlett-Packard Journal* (October 1966), pp. 12–15.

Hamilton, S. and R. Hall. "Shunt-Mode Harmonic Generation Using Step Recovery Diodes." *Microwave Journal* (April 1967), pp. 69–78.

Hewlett-Packard: from a 1967 applications note that is no longer in print.

Hewlett-Packard. *Microwave Integrated Products Catalog*, 1979. Hewlett-Packard Components, 350 West Trimble Road, San Jose, CA 95131. Fig. 9, p. 61.

Herbert, C., and J. Chernega. "Broadband Varactor Tuning of Transistor Oscillators." *Microwaves* (March 1967), pp. 28–32.

Howe, D. A. *Frequency Domain Stability Measurements: A Tutorial Introduction*, NBS Technical Note 679, March 1976.

"How to Select Varactors for Harmonic Generation." *Micronotes*, Vol. 10, No. 1 (May 1973). Microwave Associates. Inc., Burlington, MA.

Hutchinson, B. H. Jr. "Contemporary Frequency Synthesis Techniques." See Gorski-Popiel above.

IBM, *Continuous System Modeling Program III (CSMP III) Program Reference Manual*, 4th ed., Program Number 5734-XS9 (International Business Machines Corp., 1975).

Johnson, S. L. *et al.* "Noise Spectrum Characteristics of Low-Noise Microwave Tubes and Solid-State Devices." *Proceedings of IEEE*, Vol. 54, No. 2 (February 1966), p. 260.

Kasperkovitz, W. D. "Frequency-Dividers for Ultra-High Frequencies." *Philips Technical Review* (Netherlands), Vol. 38, No. 2 (1978–79), pp. 54–68.

Kincaid, R. "Basic Program Designs Versatile Linearizer Circuit." *EDN* (November 15, 1979).

Klapper, J., and J. Frankle. *Phase-Locked and Frequency Feedback Systems*. New York, NY: John Wiley, 1972.

Krishnan, S. "Diode Phase Detectors." *Electronic and Radio Engineer* (February, 1959), pp. 45–50.

Kroupa, Venceslav. *Frequency Synthesis Theory Design and Applications.* New York, NY: John Wiley, 1973.

Kurokawa, Kaneyuki. "Injection Locking of Microwave Solid-State Oscillators." *Proceedings of IEEE*, Vol. 61, No. 10 (October 1973), pp. 1386–1410.

Kurtz, Stephan R. "Mixers as Phase Detectors." *Tech-Notes*, Vol. 5, No. 1 (January/February, 1978). Watkins-Johnson Co., Palo Alto, CA.

Kurtz, Stephan R. "Specifying Mixers as Phase Detectors." *Microwaves* (January 1978), pp. 80–87.

Lance, A. L., W. D. Seal, F. G. Mendoze, and N. W. Hudson. "Auto-mating Phase Noise Measurements in the Frequency Domain." *Proceedings of the 31st Annual Frequency Control Symposium* (1–3 June, 1977). Washington D.C.: Electronic Industries Association, pp. 347–358.

Lee. S. *Digital Circuits and Logic Design.* Englewood Cliffs, N.J.: Prentice-Hall, 1976.

Leeson, D. B. "A Simple Model of Feedback Oscillator Noise Spectrum." *Proceedings of the IEEE*, Vol. 54, No. 2 (February 1966), pp. 329–330. *NOTE* The symbols ϕ and $\dot{\phi}$ are interchanged several times in this paper.

Leeson, D. B. "Short-Term Stable Microwave Sources." *Microwave Journal*, Vol. 13, No. 6 (June 1970), pp. 59–69.

Lindsey, W. C. and M. K. Simon. *Phase-Locked Loops and Their Applications*, New York, NY: IEEE Press, 1978.

Manassewitsch, Vadim. *Frequency Synthesizers Theory and Design.* New York, NY: John Wiley, 1976.

Mengali, U. "Acquisition Behavior of Generalized Tracking Systems in the Absence of Noise." *IEEE Transactions on Communications*, COM-21 (July 1973), pp. 820–826. (From Lindsey, pp. 159–165.)

Miller, R. L. "Fractional-Frequency Generators Utilizing Regenerative Modulation." *Proceedings of the IRE* (July 1939), pp. 446–457.

Morrison, R. *Grounding and Shielding Techniques in Instrumentation.* New York, NY: John Wiley, 1967.

Motchenbacher, C. D., and F. C. Fitchen. *Low-Noise Electronic Design.* New York, NY: John Wiley, 1973.

Motorola Data Sheet, MC12012, 1973. Motorola Semiconductor Products, Inc., Phoenix, AZ 85036.

Motorola Data Sheet: "Phase-Frequency Detector, MC4344, MC4044," 20 p.

Napier, R. and C. Foster. "18–40 GHz Broadband Frequency Synthesizer Technique." *Microwave Journal* (April 1977), pp. 40, 42, 43, 46, and 47.

Nicholds, J and C. Shinn. "Pulse Swallowing." *EDN* (October 1, 1970), pp. 39–42.

Ondria, John G. "A Microwave System for Measurements of AM and FM Noise Spectra." *IEEE Transactions on Microwave Theory and Techniques*, MTT-16, No. 9 (September 1968), pp. 767–781.

Oropeza, F. and J. P. Schoenberg. "Binary Frequency Synthesis . . . ." *Frequency* (Sept/Oct. 1976), pp. 14–17.

Ott, H. *Noise Reduction Techniques in Electronic Systems.* New York, NY: John Wiley, 1976.

Owen, D. P. "Measurement of Phase Noise in Signal Generators." *Marconi Instrumentation*, Vol. 15, No. 6 (Autumn 1977), pp. 117–122.

Papaiech, R. and R. Coe. "New Technique Yields Superior Frequency Synthesis at Lower Cost." *EDN* (October 20, 1975), pp. 73–79.

Parker, T. E. "SAW Controlled Oscillators." *Microwave Journal* (October 1978) pp. 66 and 67.

Penfield, P. and R. P. Rafuse. *Varactor Applications*. Cambridge, MA.: The MIT Press, 1962, Chap. 9.

Pritchard, R. L. *Electrical Characteristics of Transistors*. New York, NY: McGraw-Hill, 1967, p. 551.

Protonotarios, E. N. "Pull-in Performance of a Piecewise Linear Phase-Locked Loop." *IEEE Transactions on Aerospace and Electronic Systems*, Vol. AES-5, No. 3 (May 1969), pp. 376–386.

Racal *Technical Manual RA6790, HF Receiver RCI 84244*, Racal Communications, Inc., 5 Research Place, Rockville, MD 20850, June 1979, pp. 4-11–4-22.

Ragazzini, J and G. Franklin. *Sample-Data Control Systems*, New York, NY: McGraw-Hill, 1958.

Rey, T. J. "Automatic Phase Control, Theory and Design." *Proceedings of the IRE* (October, 1960), pp. 1760–1771. Corrections appear in *Proceedings of the IRE* (March 1961), p. 590. (Also available in Lindsey, pp. 309–320.)

Reynolds, C. "Measure Phase Noise." *Electronic Design* (February 15, 1977), pp. 106–108.

Richman, D. "Color-Carrier Reference Phase Synchronization Accuracy in NTSC Color Television." *Proceedings of the IRE* (January 1954), p. 125.

Richman, D. "The DC Quadricorrelator: A Two-Mode Synchronization System." *Proceedings of the IRE* (January 1954), pp. 288–299.

Robson, R. G. "The Pull-in Range of a Phase-locked Loop." *Conference on Frequency Generation and Control for Radio Systems: London*, Conference Publication No. 31 (May 1967), pp. 139–143. (From Kroupa, p. 177.)

Sherwin, J. "Cut Transients in FET Analog Switches." *Electronic Design* (April 27, 172), pp. 50–54.

Shoaf, J. H., D. Halford, and A. S. Risley. *Frequency Stability Specifications and Measurement: High Frequency and Microwave Signals*, NBS Technical Note 632 (January 1973).

Solid-State Microwave Voltage Controlled Oscillators. Frequency Sources, Inc., Chelmsford, MA 01824, 1974.

SP8750-8752 Data Sheets. Plessey Semiconductors, 1674 McGaw Ave., Irvine, CA 92714.

Speckhart, Frank H., and Walter L. Green. *A Guide to Using CSMP–The Continuous System Modeling Program*. Englewood Cliffs, NJ: Prentice-Hall, 1976.

Taub, H. and D. L. Schilling, *Principles of Communication Systems*. New York, NY: McGraw-Hill, 1971, pp. 235–319.

Tierney, J. "Digital Frequency Synthesizers." See Gorski-Popiel above.

Tipon, P. G. "New Microwave-Frequency Synthesizers That Exhibit Broad Bandwidths and Increased Spectral Purity." *IEEE Transactions on Microwave Theory and Techniques*, Vol. MTT-22 (December 1974), p. 1251.

Truxal, J. *Automatic Feedback Control System Synthesis*. New York, NY: McGraw-Hill, 1955, pp. 38–41.

Underhill, M. J. *et al.* "A General Purpose LSI Frequency Synthesizer System." *Proceedings of the 32nd Annual Symposium on Frequency Control*, 1978, pp. 366 and 367.

Van Duzer, V. "A 0-50 Mc Frequency Synthesizer with Excellent Stability, ..." *Hewlett-Packard Journal*, Vol. 15, No. 9 (May 1964), pp. 1–6.

Viterbi, A. *Principles of Coherent Communication*. New York, NY: McGraw-Hill, 1966.

Walston, J. and J. Miller, (ed.). *Transistor Circuit Design*. New York, NY: McGraw-Hill, 1963, pp. 307–320.

Weaver, C. S. "A New Approach to the Linear Design and Analysis of Phase-Locked Loops," *IRE Transactions on Space Electronics and Telemetry*, SET-5 (Dec. 1959), pp. 166–178. (From Lindsey. pp. 107–119.)

Westman, H. P. (ed.). *Reference Data for Radio Engineers*, 5th ed. New York, NY: Howard Sams and Co., Inc., 1968.

White, D. *Electromagnetic Interference and Compatability*, Vol. 3. Don White Consultants Inc., 14800 Springfield Rd., Germantown, MA 20267, 1973, pp. 4.1–8.30 and 10.1–12.14.

Index